John Emsley

Better Looking, Better Living, Better Loving

1807 – 2007 Knowledge for Generations

Each generation has its unique needs and aspirations. When Charles Wiley first opened his small printing shop in lower Manhattan in 1807, it was a generation of boundless potential searching for an identity. And we were there, helping to define a new American literary tradition. Over half a century later, in the midst of the Second Industrial Revolution, it was a generation focused on building the future. Once again, we were there, supplying the critical scientific, technical, and engineering knowledge that helped frame the world. Throughout the 20th Century, and into the new millennium, nations began to reach out beyond their own borders and a new international community was born. Wiley was there, expanding its operations around the world to enable a global exchange of ideas, opinions, and know-how.

For 200 years, Wiley has been an integral part of each generation's journey, enabling the flow of information and understanding necessary to meet their needs and fulfill their aspirations. Today, bold new technologies are changing the way we live and learn. Wiley will be there, providing you the must-have knowledge you need to imagine new worlds, new possibilities, and new opportunities.

Generations come and go, but you can always count on Wiley to provide you the knowledge you need, when and where you need it!

William J. Pesce
President and Chief Executive Officer

Peter Booth Wiley
Chairman of the Board

John Emsley

Better Looking, Better Living, Better Loving

How Chemistry can Help You Achieve Life's Goals

WILEY-VCH Verlag GmbH & Co. KGaA

The Author

Dr. John Emsley
Alameda Lodge
23A Alameda Road
Ampthill, Bedfordshire, MK45 2LA
United Kingdom

Library of Congress Card No.:
applied for

British Library Cataloguing-in-Publication Data
A catalogue record for this book is available from the British Library.

Bibliographic information published by the Deutsche Nationalbibliothek
Die Deutsche Nationalbibliothek lists this publication in the Deutsche Nationalbibliografie; detailed bibliographic data are available in the Internet at http://dnb.d-nb.de.

Typesetting TypoDesign Hecker GmbH, Leimen
Printing Ebner & Spiegel GmbH, Ulm
Binding Ebner & Spiegel GmbH, Ulm
Cover Design Himmelfarb Grafik und Webdesign, Schwetzingen
www.himmelfarb.de
Wiley Bicentennial Logo Richard J. Pacifico

Printed in the Federal Republic of Germany

Printed on acid-free paper

ISBN: 978-3-527-31863-6

Contents

Explaining the Things We Use Now – and News from the Future *IX*

Acknowledgements *XI*

Things to Bear in Mind *1*
- Food Calories *1*
- Organic *1*
- Trade Names *1*
- Money *2*
- Large and Small Measures *2*
- Chemical Names *2*
- News from the Future *3*
- Glossary *3*

Introduction – The Right Chemistry *5*

1 Better Looking (I): Hair, Eyes, Teeth, Nails *9*
- News from the Future *9*
- Crowning Glory *10*
- Thinning on Top *15*
- I Can See Clearly Now ... *20*
- A Gleaming Smile *25*
- Nailed *31*

2 Better Looking (II) and Better Living (I): Medical Advances *37*
- News from the Future *37*
- Skin Diseases: Acne, Eczema, and Psoriasis *38*

Better Looking, Better Living, Better Loving. John Emsley

ISBN 978-3-527-31863-6

Acne 40
Eczema 42
Psoriasis 44
An Agile or Fragile Old Age? 46
Carbohydrate as Cures 54
Anaesthetics 60

3 **Better Loving: The Not-so-dirty Weekend** 67
News from the Future 67
Smelly Chemistry 68
Wash All My Cares Away 72
It's the Pits, Man 75
Antiperspirants 78
Deodorants 81
Something for the Weekend, Sir? 83
We Just Got Carried Away Mum 88

4 **Better Living (II): Improving Our Diet** 93
News from the Future 93
Carbohydrates 95
Manna from Heaven ... 101
Low I_2 Means Low IQ 105
Functional Foods 110
Crap Food? Feeding the Fifty Billion 114
Hot, Hot, Hot 118

5 **Better Living (III): Minor Metals for Major Advances** 125
News from the Future 125
Solar Panels Provide Photovoltaic (PV) Power 126
Glass is Green 137
The Magic of Titanium 141
The Future's Blue 147

6 Better Living (IV): Chemistry in the Home *153*

News from the Future *153*
Wash All My Troubles Away: Laundry Aids *154*
Surfactants *159*
Foam Regulators *161*
Builders *161*
Anti-redeposition Agents *163*
Dye Transfer Inhibitors *164*
Peroxide Bleach and Bleach Activators *164*
Fluorescing Agents *166*
Fragrances *167*
Fabric Softeners *167*
All Washed Up? *169*
Malodours and Air Fresheners *172*

7 Better Looking (II): The Art of the Chemist *177*

News from the Future *177*
Colour *179*
Red *180*
Blue *184*
Yellow *186*
Green *187*
Purple *188*
Brown *188*
Black *189*
White *189*
Oils and Varnishes *190*
Analysis *191*
Conservation *194*
Restoration *197*
Frauds and Fakes *200*

Glossary *205*

Index *217*

Explaining the Things We Use Now – and News from the Future

Better Looking, Better Living, Better Loving is about grooming, health, food, and love making, and the products we buy to improve these areas of our life. But what do these products contain? How do they work? Are the chemicals they contain safe? Award-winning popular science writer John Emsley answers these questions using language that we can all understand – and he also asks some questions of his own. Do they come from sustainable resources? Can better ones be developed? Are those extracted from natural sources really better than synthetic ones?

There are also chapters dealing with recent developments such as new drugs to treat skin diseases, new materials for turning sunlight into electricity, new products for cleaning, and new ways of restoring famous works of art to their original state. Each chapter begins with a fictional news item from the future speculating on the possible benefits that might come as more research is carried out, and each chapter ends by addressing an issue of concern related to the topics discussed.

Better Looking, Better Living, Better Loving. John Emsley

ISBN 978-3-527-31863-6

Acknowledgements

Several people kindly agreed to read sections of this book to ensure that I had got the science and emphasis right. To all of them I owe a deep debt of gratitude. They were:

- Dr Christopher Flower, Director-General of the Cosmetic, Toiletry & Perfumery Association;
- Dr Howard Hill of Huntingdon Life Sciences;
- Dr Garry Rumbles, Principal Research Scientist at the National Renewable Energy Laboratory at Golden, Colorado;
- Dave Yost;
- Helen Glyn-Davies, dietitian at the Luton and Dunstable Hospital;
- Professor Tim Jones, Director of the Centre for Electronic Materials and Devices of Imperial College London;
- Anthony Lipmann of the Minor Metals Trade Association;
- Fiona Steele;
- Dr Alan Bailey of the Analytical Services Centre of the Forensic Science Service, London.
- Dan Lewis, Research Director of the Economic Research Council and a journalist and broadcaster specialising in energy issues;
- Alan Phenix at the Getty Conservation Institute in California;
- Professor Steve Ley and Rose Ley of the Department of Chemistry, University of Cambridge;
- My association with the UK's Broadcast Advertising Clearance Centre, which vets all television advertising, means I am well acquainted with the products discussed in Chapter 6, namely detergents, fabric softeners, dishwash tablets, and air fresheners, but I prevailed on some of my contacts in the industry to review what I had written. They, however, wish to remain anonymous but they know who they are, and my thanks go out to them also.

Better Looking, Better Living, Better Loving. John Emsley

ISBN 978-3-527-31863-6

As always, my wife Joan took on the role of general reader and read the whole book, pointing out sections where I had allowed technical details to obscure the message.

Things to Bear in Mind

Food Calories

When it comes to talking about the energy requirements of the human body we speak of calories, although we should really use the term kilocalories. A calorie will raise the temperature of a gram of water by one degree centigrade, which is rather a small quantity, whereas a kilocalorie will raise the temperature of a kilogram of water (= 1 litre) one degree centigrade and clearly is more meaningful. To say that an apple provides 40 calories therefore is technically incorrect but this is now the language of everyday speech. So that there is no confusion, I will use the term *food-calories* but remember these are really kilocalories.

Organic

This word has different meanings but in this book I will use it only in the chemical sense, so that when I refer to an organic molecule or an organic solvent I mean one that is based on carbon. There should be no confusion between this chemical use of the word and its popular use which rather oddly has come to mean things that do not include any synthetic chemicals.

Trade Names

Several products are referred to by their trade names because these are the ones that are most likely to be recognised. I acknowledge the rights of the owners of these trade names. The products mentioned in the text are Acuvue, Actimel, Advance, Acuvue-1-day, Alopexil, Alpha-

Better Looking, Better Living, Better Loving. John Emsley

ISBN 978-3-527-31863-6

derm, Avanti, Betnovate, Canesten, C-Film, Ciba Vision, Clearasil, Courtelle, Dermovate, Durex, Duron, Enbrel, Encare. Exterol, Febreze, Fiber K, Finish, Focus-1-2-week, Gaio, Grecian 2000, Humira, Hyperol, Intercept, Lonolox, Lycra, Myocrisin, Naturalamb, Noxer, Nujol, Orlon, Perhydrit, PowerGlaz, Prexidil, Prexige, Propecia, Proscar, Regaine, Remicade, Ridaura, Rogaine, Rohypnol, Sensation, Sensodyne, Spandex, Vanish, Vaseline, Vioxx, Vistakon, Yakult.

Money

Most readers will have euros (€), dollars ($), or pounds (£), as their unit of currency. The exchange rates of these vary from day to day which makes converting from one to the other an inexact science. At the time of writing (1 July 2006) €1 was equal to £0.7 and $1.2.

Large and Small Measures

As things get smaller we indicate this in jumps of 1000 by a prefix to the name. Thus a thousandth of a gram is a milligram, a thousandth of a milligram is a microgram, and a thousandth of a microgram is a nanogram. In effect a microgram is a millionth of a gram and a nanogram is a billionth of a gram.

Similar changes to names occur as things get larger, so that a kilogram is a thousand times larger than a gram, and we could talk of megagrams as a thousand times larger than a kilogram, but we prefer instead to call such a quantity a tonne. For really large amounts there are megatonnes (= 1 million tonnes) and on a planetary scale we talk of gigatonnes which are billions of tonnes. See also →Units of Measure in the Glossary for more details.

Chemical Names

There are strict rules for naming chemical compounds but these are meaningless to all but a chemist. Many substances have a common name that people recognise and I have chosen to use these names, because my aim is to make the text accessible to everyone. For readers who know some chemistry then I have added footnotes giving the alternative names as well as basic chemical formulae.

News from the Future

As far as I am aware there is no such publication as *Global Times News* – at least a search of the Internet did not reveal one – but maybe in 20 years there will be something that has a global edition which will be accessible and downloadable on to personal screens. I have started every chapter with a page from the 21 March 2025 edition (the first day of spring) giving news of a better world that I hope we can achieve, or at least of some part of that world where new benefits of chemistry will be available. I have also included other news from the future within chapters where I feel certain that such benefits are forthcoming. In some cases I have also invented names for individuals to indicate that they come from mixed ethnic backgrounds. My purpose in doing so is a belief in a world of the future when racial and religious conflicts disappear as we discover our common humanity and work together to achieve a sustainable and equitable future for all species, for future generations, and for planet Earth. Clearly such a world will take far longer to achieve than 2025 although the benefits that these news items are describing could well be with us by then.

Glossary

This provides extra information about subjects that are preceded by a small arrow (→) in the main text but it does assume a certain level of scientific knowledge on the part of the reader.

Introduction – The Right Chemistry

This book is for those who would like to know more about some of the things that they encounter in their everyday life and which are produced by the chemical and pharmaceutical industries. In *Better Looking, Better Living, Better Loving* my aim is to explain what these are and why they work as they do. I am aware that some of them have been made to appear quite worrying, unduly so in my opinion, but I will try when discussing products that have been targeted as risky to present the arguments both for and against.

We would all like to live in a world free of hunger, disease, and poverty – and maybe that's an impossible dream. We would all like to live a life that is healthy, long, safe, and enjoyable – and maybe that's an impossible dream as well. We would all like our needs to be met by using only renewable resources – and that too seems an impossible dream. In fact none of these dreams is impossible. The application of chemistry has brought great benefits to millions in the developed world and there is no reason why these benefits could not be enjoyed by all the Earth's inhabitants, and for as long as humans inhabit this planet.

Almost every week some new product is launched or some old product reformulated to offer something better. A modern supermarket is a veritable Aladdin's cave with more than 10,000 items on offer, and there are many products that have to bought and used on trust. What do they contain? The list of ingredients may tell us very little – indeed might appear somewhat threatening if we don't understand the names being used – yet for everything there should be a reason for its being included. *Better Looking, Better Living, Better Loving* will explain these reasons for some of them.

People also worry about medication, especially when they read of people's health getting worse as a result of taking a particular drug. To what extent should such headlines worry us? If an adverse reaction

Better Looking, Better Living, Better Loving. John Emsley

ISBN 978-3-527-31863-6

affects only one person in 10,000 why should we deny its benefits to the other 9,999 who are being helped by it? However, if the risk is much higher, say 1 in 100, then we might well want to avoid using such a drug, and yet we should be aware that 1 person in 65 who undergoes surgery will be made worse as a result.

Sometimes it is difficult to separate science from speculation when we read some disturbing report about chemicals. Many people fall prey to imagined dangers or perceive minor hazards to be major risks, but how can we make decisions if we are given only limited and possibly biased information? My advice is to look for *scientific* proof in support of what is being said. The hallmark of a scientific experiment is its *reproducibility*. In other words, if an experiment is performed or the data analysed by another person in another laboratory they will reach the same conclusion. Of course nothing is entirely reproducible and small deviations have to be allowed for, and it is accepted that small experimental errors will creep in because conditions are never exactly the same, so we have to rely on averages and note that there is a degree of uncertainty which scientists can calculate in order to assess the reliability of the information. Some people would like scientists to give assurances that something is *absolutely* safe but of course they cannot do so.

We should always suspect news stories that present information in an alarming way. Some are merely manufactured news and an example of this is the finding of unnatural chemicals in the human body. This can be made newsworthy if the body is that of someone with a high profile such as a politician, actor, or media personality. What those who carry out such tests fail to tell you is that these chemicals are there in amounts so tiny they can do no harm. They may even be detectable at levels of parts per trillion (p.p.t.) an amount so small that it is like comparing one *second* to 30,000 years. There are other quite natural components of our body that might worry you. Did you to know that each one of us contains a *trillion* atoms of uranium? That's rather a frightening way of presenting data. It is based on a figure that is much less frightening, which is that the average person contains only a few micrograms of this metal (a speck of dust weighs a microgram). This uranium comes from the food we eat, which in turn comes from the soil, where traces of uranium occur perfectly naturally.

We all want to lead a green lifestyle but how do we reconcile this with our desire to benefit from the things the chemical industry produces and which we use every day. One of my aims in writing *Better Looking, Better Living, Better Loving* was to try and show that the two are not irreconcilable, and that chemical industry can move towards sustainability and indeed there is no reason why my grandchildren cannot enjoy all the wonderful products I enjoy, and that their grandchildren in the next century will do so as well.

Welcome on board my tour of some of the wonders of Chemistry City. We will be visiting the cosmetic factory, the pharmacy, the grooming salon, the diet clinic, the power plant, the domestic cleaning company, and end at the art gallery. If you want to debate an issue of concern then there is a section at the end of each chapter where you can linger. For those of you who want to know about what is to come this century, I begin each chapter with a cutting from a newspaper of the future to whet your appetite.

So fasten your seatbelts, because some of you may be in for a bumpy ride as we drive over some popular misconceptions about chemicals.

[Ampthill, July 2006]

1
Better Looking (I): Hair, Eyes, Teeth, Nails

A small arrow printed before a word in the main text indicates that there is more information on that topic in the Glossary.

News from the Future

Global Times News, 21 March 2025

Dentist Numbers Fall Again

For the fifth year running the number of registered dentists in Europe has declined and in some areas people are travelling up to 50 kilometres to obtain their services. The drop in numbers is blamed on the success of the new toothpastes such as LoveSmile that were launched 15 years ago. These not only keep teeth brilliantly white but contain nanoparticles which can penetrate cavities and repair them from the inside, so there is no longer a need for dentists to drill and fill.

"Dental work now mainly consists of fitting braces and occasionally mending broken teeth," said a leading professor at Rome's Dental School, adding: "In my early days as a teacher I spent most of my time instructing students on how to repair or replace decayed teeth. Today a call for these services is almost unheard of. Even cosmetic treatment such as tooth whitening is rarely required."

Modern toothpastes, first marketed in 2010, have been responsible for young people in Europe today having such wonderful teeth. They contained not only the usual cleaning agents and fluoride but included repairing additives such as nanoparticles of hydroxyapatite, the natural chemical from which tooth enamel is made. It is these which penetrate any tiny cavities and repair them. Products like LoveSmile also contain a whitening agent which prevents staining and, unlike the earlier whitening agents, it does not cause thinning of the outer layer of tooth enamel.

Page 3: Chemicals in toothpaste could cause bone cancer in old age warns dental fillings manufacturer.

Of course there is no such product as LoveSmile toothpaste, at least a search of the Internet failed to reveal one. What the above news item suggests is that a trip to the dentist is likely to become a rare event in the life of children born today, unlike the regular visits to the dentist that most people have had to make in the past if they wanted to keep their teeth in good condition. The new ingredients mentioned in the

Better Looking, Better Living, Better Loving. John Emsley

ISBN 978-3-527-31863-6

news item are already known and we will look at them later in this chapter, the theme of which is improving our looks with the help of various products that chemists have devised, namely hair colorants, hair restorers, contact lenses, tooth whiteners, and artificial fingernails. We begin with hair, regarded by many as their best asset, but it is possible to make it even more attractive.

Crowning Glory

Hair comes in various natural shades and colours, ranging from jet black through brown and auburn to ash blonde, which are produced by varying amounts of two pigments. These are slightly different versions of a biological →polymer called melanin, one is black, and called eumelanin, and the other is blonde, and called phaeomelanin. Eumelanin particles predominate in black and brown hair, while phaeomelanin particles dominate in fair hair. These melanin pigments are incorporated into the hair as it grows, and the particles are formed in special cells called melanocytes. (When these cease to function then the hair they produce has no colour.) What our genes provide may not be quite what we want, and so we may seek to change the colour of our hair and this option is proving popular with young and old alike. Many will look to a hair stylist for advice, and they in their turn look to chemists to provide the necessary materials to make it possible. We can also purchase packs of these same chemicals and make the change ourselves.

It will come as no surprise to learn that the global market for hair dyes exceeds $7 billion annually, and it continues to grow, with the US accounting for around a quarter. The first truly permanent hair dye was created by a French chemist, Eugène Schueller in 1907 who founded the French Harmless Hair Dye Company. Today we know it as L'Oréal and it is still the biggest manufacturer of hair dyes with just over a third of world sales. Next comes Procter & Gamble with 13%, followed by Henkel with 11%, Wella with 9%, and Hoyu with 6%. Smaller producers account for the remaining 25%.

Hair dyes are of three kinds: temporary, semi-permanent, and permanent. The first kind are easily washed out of the hair, the second will survive several washings but eventually fade, and the third are locked fast within the hair shaft and disappear only as the hair grows out. What differentiates the various dyes is their chemistry, and it is the permanent type which is the most sophisticated chemically, although some people find this worrying – more of that in a minute. First the good news.

The dyes that are used for temporary colouring can be applied in the form of rinses, gels, mousses, and sprays. They coat the surface of the hair with dye and most of this will be washed away the next time the hair is shampooed. We may want this to occur if we have applied the dye for a special event such as a party, a fun day, or an on-stage performance. A typical temporary hair dye is the →colorant known as FD&C Blue No.1.[1] This is a large molecule so it cannot penetrate into the hair, and it has three negatively charged groups of atoms which make it highly water soluble so it is easily washed off.

Semi-permanent dyes are smaller molecules and these can pass through the outer layer of the hair, known as the cuticle, and move into the inner cortex, there to remain until they gradually leach out again and are washed away. A combination of dyes is generally required to produce the desired shade and those commonly used are listed in the Glossary.

Chemicals destined to produce permanent hair dyes also penetrate the cuticle but then they become trapped by reacting together to form a much larger molecule which cannot escape. For this to happen the cuticle must be made permeable, which can be achieved by the action of a little ammonia (NH_3). This causes the hair shaft to swell by raising the →pH to about 10, and this opens up the cuticle scales. Once inside the hair, the molecules react to form the dye molecule which is too large to escape when the scales close again under the action of a final rinse. Permanent hair colorants come as two separate gels which have to be mixed together before application. One contains hydrogen peroxide which serves two functions, one is to bleach away the natural melanins, the other is to activate PPD (short for paraphenylenediamine[2]) which is then combined with a second molecule, known as a coupler, to form the dye. In fact there may be more than one type of coupler in a hair dye. For example, in L'Oreal's dye 'Havana' there are three couplers[3] which together generate a pale auburn shade referred to as light amber.

1) FD&C refers to the US Federal Food, Drugs, and Cosmetics Act.

2) Also called 1,4-benzenediamine.

3) These are 4-amino-phenol, 4-amino-2-hydroxy-toluene, and 3-amino-phenol.

That was the good news ... now the bad news:
There is another type of hair colorant which is generally aimed at men who are going grey and is based on lead acetate. Although it is poisonous it is regarded as safe because the metal is not absorbed through the skin. The dye is rubbed on the hair with which it reacts to form a black pigment. With each application the hair gets darker and darker until all the grey hairs have disappeared. Such treatments date back to the days of the Roman Empire, when combs made of lead were used and were dipped in vinegar. The acetic acid of the vinegar dissolved a little of the lead, which was then transferred to the hair. For the past 200 years or so the preferred treatment has been to dab a solution of lead acetate directly onto the hair, and such products are still available. Soon they are to be banned inside the EU on the grounds that lead is inherently dangerous.

More serious perhaps are the charges that permanent hair dyes may cause cancer, they have even killed those who used them. Although PPD has been an ingredient of hair dyes for more than 30 years it has not been without occasional bad publicity due to an allergic reaction in some people. We now know that this type of allergy affects only three women in a million, and it has been traced to a rare genetic susceptibility. Many years ago it was claimed that as many as one woman in a hundred was sensitive to PPD. This turned out to be somewhat of an exaggeration and came from tests in which pure PPD was applied directly to the skin and then covered with a plastic patch. Nevertheless, users of these dyes are warned that a skin allergy test must be carried out. Typically they are told to apply a penny-sized area of the dye to the skin behind the ear. When this has dried, a second coating of dye should be applied, and then left for 48 hours. If the skin becomes inflamed in any way then the dye must not be used.

In May 2001 a Mrs Narinder Devi who lived in Birmingham, England, decided to dye her hair but she skipped the skin test. Sadly she suffered a massive anaphylactic shock from which she died. This was a rare example of a fatality following the use of a hair dye, although what ingredient caused it was never deduced. Other severe cases of allergic reaction continue to occur, although rarely with such a fatal outcome, and while these are widely publicised in the media, they are extremely rare. (Victims invariably settle out of court.) Of women who genuinely do test their skin for sensitivity, around 2% will observe a slight positive effect and they are informed in the leaflet which comes with the hair dye that they must not use the product.

The 1970s also saw claims that the various chemicals used in hair colorants caused cancers in laboratory animal tests when fed to them in large doses over a long period. Others reported that they caused mutations in bacteria. As a result the suspect ingredients were withdrawn. Although PPD was never implicated in these tests, the EU even introduced limits on PPD to the effect that no more than 6% can be present in hair colorant products. Tests showed that in a typical hair dyeing session a women would absorb some PPD (at most 36 mg) but in any case this was rapidly excreted in her urine. PPD has been thoroughly tested over many years, but that does not prevent scare stories about its use being circulated.

PPD is manufactured on a large scale by the US chemical giant Du Pont and is mainly used to manufacture resins and polymers. That which is destined for inclusion in hair dyes is specially made by a different process that ensures it is absolutely pure and that it is free from any by-products that could cause adverse effects. What this means is that if there is a risk to health from exposure to PPD then it will be due to the PPD itself and not to some hidden cause. There is an alternative to PPD and that is TDS (short for toluene-2,5-diamine sulfate). This is less skin sensitive but produces shades of brown that are slightly redder.

Hair dyes have come under attack many times from those who thought their use might have long term effects such as causing cancer. In 2001 an →epidemiological report linked them to bladder cancer, while another in 2004 linked them to leukaemia. Both reports naturally attracted media interest, and continue to be quoted, despite the fact that similar, and often better conducted, studies found no such links. The leukaemia scare began with a paper in the *American Journal of Epidemiology* which reported a study of 769 adults with acute leukaemia and compared them with 623 adults who did not have leukaemia. The finding was that those who had used older-style permanent hair dyes in the 1980s were more at risk of leukaemia, although that risk was tiny. (There was no extra risk for those who used non-permanent dyes.) The paper attracted worldwide attention although it was a retrospective analysis, and the hair dyes it referred to were phased out more than 20 years ago.

A survey in 2000, carried out in Los Angeles by a group at the University of Southern California School of Medicine, found a link with bladder cancer. It involved 1541 people with this condition and they were compared to 897 people who were not so afflicted. Adjustments were made to take into account smokers, who are liable to be more at risk of this disease in any case, and it found that those who used permanent hair dyes every month had a slightly higher risk of bladder cancer, especially if they had used hair dyes continuously for 15 or more years. Hairdressers had an even higher risk. This revelation prompted the Scientific Committee on Cosmetic Products and Non-Food Products of the EU (SCCNFP) to issue a discussion paper in February 2002. This was followed in December that year by a requirement that manufacturers must submit by July 2005 all their data on hair dyes together with studies to assess their safety in terms of cancer and toxicity. The outcome of this has yet to be published, but it will take the form of an approved list of hair dyes which will be issued in 2007.

In 2004 another epidemiological study was undertaken of 459 cases of bladder cancer in New Hampshire, USA, by the Dartmouth Medical School at Lebanon, and they were matched against 665 people who did not have the disease. This study found that men who used hair dye were *less* likely to suffer bladder cancer, whereas for women there was a slightly higher risk, although in both cases the observations were not statistically significant. Another survey, this time of more than half a million women, and carried out by the American Cancer Society, found no link at all between hair dyes and bladder cancer.

A survey of 608 Connecticut women with breast cancer and 609 women who were free of the disease, was carried out under the auspices of Yale School of Public Health with collaboration from the European Institute of Oncology in Italy, McGill University in Canada, and the US National Cancer Institute. This weighty team of international medics found no evidence that those who used hair dyes of either the temporary or permanent kind were in any way increasing their risk of having breast cancer, and they published their results in the *European Journal of Cancer* in August 2002. In 2003 another study was undertaken by the world-famous Karolinska Institute of Sweden and specifically looked at the incidence of all types of cancer in Swedish hairdressers, the group that was expected to be most at risk.

The medical records of 38,800 women and 6,800 men were consulted stretching back for 40 years and these found that there was a higher risk of cancer in the 1960s but not in subsequent years, and there was no increased incidence of bladder cancer among the hairdressers compared to the rest of the population.

So what should we make of all this? The upshot of all this analysis would appear to be that modern hair dyes present no risk of causing cancer either among those who apply them or those on whose hair they are applied. If you are still not convinced that synthetic chemical dyes are safe, and yet you want to change the colour of your hair or hide any grey then you must perforce turn to the dyes of old – see box – but even some of these should come with test patches and a warning.

'Natural' Hair Dyes

A traditional hair dye is henna which is extracted from a privet-like shrub, *Lawsonia inermis*, which grows in India, Pakistan, and Egypt, and which produces the chemical lawsone. This acts as the dye molecule and its name is 2-hydroxy-1,4-naphthoquinone and it is also known as natural orange 6. Were it to be produced by a chemical company it would be banned because it would not pass today's stringent health and safety checks; it can cause allergies and asthma in some people. Moreover, as a colorant it gives unpredictable results. Not that this stops those who campaign on behalf of so-called 'green' alternatives from advocating henna as a safe natural dye. Those who recommend henna claim it has other benefits such as preventing dandruff, killing head lice, and curing ringworm, but these claims are unproven and almost certainly unreliable.

Other 'natural' hair dyes that have been used down the centuries have been indigo (chemical name 2-[1,3-dihydro-3-oxo-2H-indol-2-ylidene]-1,2-dihydro-3H-indol-3-one) extracted from *Indigofera,* a plant of the pea family, and pyrogallol (chemical name 1,2,3-trihydroxybenzene) extracted from walnut shells. This last dye was banned for use in the EU in 1992.

Lemon juice, saffron, cloves, and tea are other plant extracts used to change the colour or tint hair, but they are somewhat unreliable, and may even have no effect at all.

Thinning on Top

Few things are more eye-catching than well styled hair, and that is true for both men and women. But while a woman's hair remains a valuable asset to the end of her life, for many men it starts to disappear when they reach 30, and soon their baldness begins to send out a signal which they may not care to broadcast, especially if they still want to be seen as sexually attractive. Not that this should have bothered the Old Testament prophet Elisha but, like many men, he was surprisingly sensitive about his looks. When he was on a journey to Bethel, sometime around 850 BC, he was mocked by a group of young boys on account of his baldness, he was so upset by their remarks that he cursed them. Suddenly two she-bears emerged from a nearby wood and tore into the group, badly savaging 42 of the children – or so it says in the *Bible*.[4]

Neither cursing, nor praying, is the answer to baldness, whereas chemistry can offer a realistic chance of undoing the ravages of time. The average head has 100,000 follicles from which hair grows at the rate of 0.37 mm per day, amounting to around 14 cm per year (just under 6 inches). Every day we lose about 50 hairs and this is perfectly natural and occurs because a hair follicle enters a resting phase for a few weeks after it has been active for around three years. The old hair then falls out and a new growth phase begins, but the new hair may be somewhat different from the old hair. It may lack melanin so it appears white, or it may be thinner, in which case it is likely to be even thinner after the next resting phase, and eventually it may not even appear at all. Hair cells divide in the bulb at the bottom of the hair follicle and as they migrate upwards they deposit a layer of →keratin to form a tube-like structure. The cells divide about once a day, which is rapid for the human body, and this is why chemotherapy treatment for cancer also affects them. Cancer cells divide as quickly as hair cells, so that drugs designed to prevent cancer cells dividing also stop hair cells from working. When these cannot divide, the follicle behaves as if it is in the resting phase and the existing shaft of hair comes loose.

Male baldness manifests itself on the crown of the head and at the temples, and it will progress until most of the scalp is hairless. This loss of hair has little to do with lifestyle, but partly to do with race; baldness afflicts 50% of white males, 22% of orientals, and 18% of black males. Equally important are the man's genes which govern production of the male hormone testosterone and its conversion to its more active form di-hydro-testosterone[5] by the enzyme *5-α-reductase* which

4) *The Second Book of the Kings* 2:23.

5) As its name implies this is testosterone with two extra hydrogen atoms attached.

regulates the functioning of several parts of the male body including the genitals and the growth of hair. This enzyme eventually causes the number of active hair follicles to decline, while those that are active spend less time in the growing phase and produce hair that is thinner. One of the paradoxes of dihydrotestosterone is that it reduces the growth of hair on the scalp but promotes growth of hair on the chin, chest, and groin. Men with naturally low levels of *5-α-reductase* do not become bald as they age, and they have much less body hair than normal. Baldness is the legacy inherited from a man's parents and there used to be nothing that he could do about it, that is until the 1980s when something strange began to be reported: the hair of some bald men suddenly started to grow again.

Throughout history there have been those who have sought to exploit men who were losing their hair by selling them expensive cures, none of which worked. Hippocrates in 400 BC recommended applying a poultice of pigeon droppings, horseradish, and nettles. Today many men benefit from two accidental discoveries, or rather from the side effects of two pharmaceutical products that they were prescribed by doctors for very different ailments. Men who were given them reported that the hair on their bald heads was reappearing. These products are now widely available and they are marketed under the names of Regaine (Rogaine in the US) and Propecia.

Propecia is the hair restorer that depends for its effectiveness on the chemical finasteride, which was designed to treat swelling of the prostate, the gland that produces the fluid part of semen. Prostate enlargement results from an excess of dihydrotestosterone, and is a condition that manifests itself in many men over 65. Finasteride was devised by chemists of the drug company Merck Sharp & Dohme and the process for making it was published in 1986 in a paper in the *Journal of Medicinal Chemistry*. This also reported on the way the molecule attached itself to *5-α-reductase* and by blocking this enzyme the level of dihydrotestosterone is reduced. The link between this hormone and an enlarged prostate had been noted more than a century ago, in that castrated males never suffered from an enlarged prostate. As the prostate grows larger it squeezes the urethra, the tube down which urine exits from the bladder and this causes difficulty in urinating and the need to pass water frequently. The prostate may also develop cancer and while this is not a particularly life-threatening cancer it only makes matters worse.

The finasteride molecule has the same shape as testosterone – it has a nitrogen atom in place of a carbon atom – and it can fit into the enzyme's active site. The upshot is that the enzyme encircles the molecule thinking it has hold of testosterone but finds it cannot carry out the modification which it is supposed to perform. The enzyme clings on to the finasteride in the belief that it has the right molecule and the result is stalemate. Finasteride works mainly on the *5-α-reductase* in the genital region, but it also moderates this hormone in the scalp as well, so that hair begins to grow again. This unexpected side effect eventually led to finasteride being available under two brand names: Proscar, which is prescribed for the treatment of enlarged prostate, and Propecia, which is prescribed to treat baldness. Within a few years Proscar was generating revenues of $500 million a year in the USA alone, and Propecia was eventually to equal this. Proscar tablets are blue and apple-shaped, and the dose is 5 mg. Propecia tablets are tan coloured and octagonal in shape, and the dose is 1 mg. Those taking finasteride are warned that it must be taken for three months for it to have any effect.

Is Propecia an effective hair-restorer? The answer would appear to be yes, based on comparisons with a placebo, of which only 7% of men taking it said they noticed signs of regrowth, whereas 66% of men who were taking the real thing said there were visible signs of new hair. Interfering with a male hormone might have been expected to have another side effect, namely lowered sex drive. However, only 2% of men taking Propecia reported that this was so. Nevertheless, sexually active men who take Propecia should not risk impregnating their partner because if the fertilized egg is to be male then it may develop into a baby boy with some female characteristics. Finasteride must, of course, only be used by men, which is not the case for the other drug that acts to restore hair, and that is Regaine (Rogaine). This can be used by both men and women.

Regaine is the brand name for minoxidil, a drug produced by the UpJohn Corporation of America, and one that was designed to reduce high blood pressure, which it does very well. As with finasteride, what its users hadn't expected to experience was a regrowth of the hair on their heads. Minoxidil was patented in 1967 although clinical trials for treating baldness with it were not undertaken until the 1980s. The drug is a vasodilator, in other words it relaxes constricted blood vessels thereby allowing blood to flow more easily round the body. Doses of between 5 and 50 mg per day can be prescribed, patients generally starting on the lowest dose which is gradually increased.

Minoxidil as a prescription drug is known as Loniten and is taken as a tablet twice a day. It is used when other drugs have failed to control a patient's high blood pressure or when this is rising rapidly. Minoxidil starts to work almost immediately and within an hour the blood pressure will fall markedly. One side effect is fluid retention, most noticeable as a swelling of the ankles, so that a diuretic is prescribed at the same time to enable the body to increase its output of urine. The other side effect, if minoxidil is taken for several weeks, is hair growth around the face, which may explain why some women are reluctant to take the drug even for high blood pressure.

When minoxidil gets into the blood stream it passes a message to the lining of the blood vessels telling them to relax. It does this in several steps, ultimately leading to an opening of the channels that allow potassium ions to move through cell membranes and triggering the desired response. This is no doubt part of the way it works in hair follicles and, by increasing the blood supply there, it revitalises the formation of keratin. Why hair begins to re-grow on the heads of those taking minoxidil is still not truly understood, but it happens, and UpJohn sells the drug as a lotion containing 2% (for women) or 5% (for men).[6] It is applied directly to the scalp and needs to be used every day. It was the first drug the FDA approved for the treatment of hair loss and that was in the 1980s. Today there are several generic forms on sale, such as Alopexil, Lonolox, Prexidil, etc., and they are available as over-the-counter treatments, but Regaine is the market leader.

Regaine is reputed not only to stop hair loss, which it does for 80% of men using it, but appears even to stimulate re-growth in some cases. Applied twice daily it should begin to produce visible results within three months, although it could take twice as long, and may never work for some. The most noticeable side effect is itchiness, but this will subside after a week or so. In 1985 Regaine was tested at the Glasgow Royal Infirmary hospital in Scotland on 66 volunteers aged between 18 and 50. Half were given Regaine and half were given a placebo to rub on their balding scalps. At first the effects were disap-

6) Minoxidil is a white crystalline solid which melts at 248 °C, and while it is insoluble in water, it is perfectly soluble in propene glycol, and this is what the lotion contains.

pointing for those using Regaine until it was applied twice daily, and this is now the recommended treatment.

Somewhat perversely, it appears to those who start using Regaine that *more* of their hair is being lost. This is actually a sign that it is working and what they are seeing is hair follicles that have become dormant being activated again and the old hair displaced. Nor should they worry that the new hair is soft and downy; it becomes thicker and stronger as the treatment is continued. However, stop using Regaine and its effects will cease and its benefits will have disappeared within a few weeks. Neither it, nor Propecia, is a permanent cure for baldness but at least they offer men who become prematurely bald some hope of keeping a more youthful appearance.

I Can See Clearly Now ...

"Boys don't make passes at girls who wear glasses" went the old adage, so what did a young girl do who needed to wear spectacles? She left them at home when she went out looking for a boyfriend. Of course there was always a risk that if she found a boy she really liked, he might reject her when he eventually saw that she needed to wear glasses. Those were the days when young men could pick and choose because they were in short supply, millions having died in two world wars. Today, many men are surplus to requirements and women can be choosey, so men are the ones who don't want to be seen wearing glasses. In fact there is no need for either sex to wear them; they can wear contact lenses instead.

Contact lenses come in a wide variety: there are soft lenses and hard lenses; there are some you can wear night and day for a month or more, and disposable ones that you need wear only for a day; there are lenses that let your eyes breathe; there are bifocal lenses; and there are even lenses to correct astigmatism, a condition of the lens of the eye which makes everything seem blurred. All are products of years of research by chemical companies like Novartis, Ciba Vision, and Wesley Jessen. There are even coloured lenses that can make blue eyes bluer, and sport lenses that enable tennis players to keep their eye on the ball by enhancing the colour of a yellow tennis ball relative to its surroundings.

The *idea* of contact lenses is not new. Leonardo da Vinci suggested them as long ago as 1508, although he never got around to making any. That did not happen until 1888, when a Dr F.A. Muller of Wiesbaden, Germany, made one from glass and it was worn by a patient of his who had no eyelids. The lens preserved the sight in one eye and he is said to have used it for 20 years. That same year a Dr Adolph Frick at the Ophthalmic Clinic in Zurich fitted six patients with glass contact lenses 1.4 cm in diameter (1/2 inch) and designed to cover the whole eye. They were not successful because he could not make them a perfect fit, though he made plaster casts of the eyes of cadavers, and even of his own eyes, to use as a template. His contact lenses were just too painful to wear.

Despite the efforts of other doctors to make better impressions of eyeballs, there was no major advance in contact lens technology for another 50 years until they could be made of the →polymer PMMA, short for poly(methyl methacrylate), which had been discovered at the laboratories of Rohm & Haas in Germany in the 1920s. They called the new polymer Plexiglas but it remained a curiosity because the chemical from which it was made was too expensive. That changed when chemist John Crawford, working for ICI in England, found a way to make methyl methacrylate cheaply from acetone, and the company named its PMMA Perspex. (In the USA it was called Lucite.) PMMA soon became a most profitable polymer, and it was ideal for all sorts of things such as illuminated signs, hospital incubators, car headlights, and aircraft windows.[7] It was ideal for contact lenses.

What features should a contact lens have? Clearly it should be as transparent as glass, fit perfectly to the shape of the cornea, and be comfortable to wear. It should not harbour microbes that might cause eye infections, and it should be wearable for days without needing to be changed. It must not block off oxygen because the cornea of the eyeball needs a supply of this vital element. If it is to be disposable, it must be inexpensive. Producing a plastic with all these benefits has almost been achieved, thanks to polymer chemists. PMMA was the first major step forward. It has a refractive index like that of glass, in other words it makes a good lens because it can collect and focus light rays.

7) The Spitfire fighter planes of World War II had Perspex windows. When pilots were injured by splinters from them, surgeons noted that fragments could remain embedded within the body because this plastic is tolerated by living tissue.

It can be shaped by moulding and, equally important, bacteria find it hard to colonise.

An American, Newton Wesley, set out to make contact lenses from PMMA in 1944, and these were for his own use. He suffered badly from a swelling of the eye, but he was in the right location to do something about it because he was a faculty member of the Monroe College of Optometry at Chicago, Illinois. There he teamed up with a bright student George Jessen, and together they worked in the basement of the boarding house where Wesley lived, using a sewing machine as a lathe to shape pieces of Lucite. His PPMA contact lenses were made to cover the whole eye and they were a success. In 1949, he and Jessen began to teach other lens technicians how to make them, and optometrists how to fit and adjust them. By 1955 their company, Wesley-Jessen, was a success and they were spending $500,000 a year on advertising. They also engaged in long-term research, regularly checking 350 of their customers who wore their lens. Meanwhile a Kevin Tuohy was working on an even simpler idea: that the contact lens need only cover the cornea of the eye and he patented such a one in June 1960. His lenses were a better fit, were more comfortable to wear and, what was most unexpected, they stayed in place and did not move around the eye ball as one might have expected.

Contact lenses made from PPMA are now history. The reason is that they deprive the cornea of oxygen, which it must get directly from the air because the cornea has no blood vessels, and the lack of oxygen can eventually cause damage. What replaced PPMA was another transparent plastic that was gas permeable and which had been discovered in Prague, Czechoslovakia, in the early 1950s. Polymer chemists Otto Wichterle and Drahoslav Lim had modified PPMA by attaching water-attracting groups to the polymer chain. The new material was called HEMA (short for hydroxyethyl methyl methacrylate) and they had originally intended using the polymer to make artificial blood vessels, but when Wichterle dislodged some which had congealed at the bottom of a test-tube he noticed how like a contact lens it was. He had inadvertently made the first soft lens. HEMA is known as a hydrogel, which means that it is a substance that attracts water and holds it in a framework of polymer molecules.

Although its softness was a real benefit, HEMA still did not allow significantly more oxygen to penetrate through to the cornea. Nevertheless the lenses, marketed by Bausch and Lomb under the brand name Soflens, were put on the market in 1971 and were an instant hit. The drawback was that they could harbour germs and so they needed to be cleaned every evening, ideally with →hydrogen peroxide, and treated once a week with an enzyme solution of either papain (extracted from pineapple) or pancreatin (from pork) to remove the fat and protein residues which collect from the tears that lubricate and protect the eyes. These residues deposit on the surface of a lens thereby allowing microbes to grow.

Polymer chemists also reformulated HEMA by adding measured amounts of other polymer precursors and then polymerising the mixture to form what is known as a copolymer. Varying ratios of copolymers were tried until a hydrogel was obtained which had all the right features, and in particular a negatively charged surface, so that the contact lens would cling to the film of tears on the surface of the eye. Contact lens wearers now have a choice of improved hydrogel types, and while these are better than the older forms they still rely on water to transport oxygen to the eye. Some have as much as 75% water content and this needs to be high if the lenses are to be worn for extended periods.

The problem of making soft lenses 'breathable' was eventually solved by attaching silicone groups to PPMA polymer. Silicones dissolve oxygen very easily, but add too many silicones and the polymer becomes water-repellent, something we might expect because silicones are often used to make water-proofing materials. These new lenses were patented in 1974 and marketed in 1979. They were referred to as rigid, gas-permeable lenses. The softness and comfort of the older HEMA lenses had been sacrificed in order to prevent long-term damage to the eyes and they had to be custom made for each individual if they were not to be uncomfortable to wear. Softer hydrogel contact lenses, which also incorporated silicone, became available in 1998 and were immediately popular with sales exceeding $150 million within five years. The trouble was that they sometimes glued themselves to the eyeball because they tended to suck out lipid molecules from the cornea and these acted as an adhesive.

How did the chemists of companies like Johnson & Johnson and Novartis achieve the seemingly impossible task of combining polymers that appear to be irreconcilable? HEMA attracts water but does not take up oxygen very well while silicones repel water but absorb any amount of oxygen gas. At first the combination of silicone and hydrogel gave only an opaque material although that was solved by incorporating nanosized components that are smaller than the wavelengths of →light so that they appeared transparent. Today there are various silicone-hydrogel materials on the market with water contents ranging from 25–45% and all having excellent oxygen absorption and transmission.[8] Some contact lenses incorporate a fluoroether, another chemical which is particularly good at absorbing oxygen.

Silicone-hydrogel contact lenses can even be worn when asleep, but there are some drawbacks to them, the main one being that they are stiffer, although one type, Acuvue Advance, is only marginally stiffer and stiffness need not necessarily be a problem because increased rigidity makes handling the lenses easier. Another manufacturer, Johnson & Johnson, has minimised stiffness by reducing the amount of silicone and putting a layer of PVP, short for poly(vinyl pyrrolidone), as a 'wetting' agent on to the surface. This PVP layer has been added to overcome one of the major drawbacks of soft lenses which is that they can dry up the eye, as many as half their users reporting this discomfort, which is sufficiently irritating to cause many of them to discontinue wearing them. Another way to counteract dryness is to expose the lenses to a gas plasma which creates a permanent ultrathin layer of silicate on their surface, this being formed from the silicone. The resulting silicate does not attract lipids and has increased wettability.

Acuvue Advance also blocks out →ultraviolet rays (UV), filtering out more than 90% of UVA, and 99% of the UVB rays that are more dangerous to the eye. (These lenses are no substitute for protective goggles which some workers wear to screen their eyes from UV light.) Some disposable contact lenses, such as Acuvue-1-day, are meant to be changed every day, while some, such as Focus1–2 week, produced by Ciba Vision, last longer. The daily ritual of cleaning contact lenses puts many off from wearing them but there are some that can safely be left in the eye for a week, and there are some experimental ones that have been left as long as three months. Ted Reid of Texas Tech University Health Science Centre in Lubbock ,Texas, has found that this becomes possible if the lenses are coated with a selenium compound, and the coating need only be one molecule thick. The coating binds itself

8) Nathan Ravi of Washington School of Medicine in St Louis, Missouri, has developed a hydrogel that can be injected into the eye, there to form a new lens to replace a diseased or ageing one. The research is as yet only in its early stages.

chemically to the lens and tests have shown that it can remain in place for as long as two years. These lenses have yet to be approved by the FDA for sale to the public.

A Gleaming Smile

You may be looking good, thanks to your hair, and seeing better, thanks to contact lenses, but when you smile at the boy or girl of your dreams, are you spoiling your chances by having discoloured teeth? If you suspect that this may be so, then chemistry can help.

Unsightly teeth or not, you should think yourself lucky that you are part of this generation. Much earlier generations had a hard time with their teeth as we can see from their remains. A few people in antiquity had the benefit of dental care because we know from mummies that there were dentists in ancient Egypt 4000 years ago. Although they did little more than pull rotten teeth, they sometimes plugged the hole left behind with gold to which they attached a false tooth made of ivory. Their skills were passed on to later civilizations such as the Etruscans and Romans, who developed more sophisticated techniques. Etruscan dentists of the sixth century BC were the first to construct gold bridgework to which they attached artificial teeth made of bone or ivory to fill the gaps left by extracted teeth. Not only did these look good but they clearly were strong enough to eat with. Dentistry in the West declined in the Dark Ages (500–1000 AD) and did not really emerge again until tooth decay and toothache became a major problem from the 1600s onwards, due mainly to cheap sugar imported from the plantations of the New World. Sugar can be converted to acid by bacteria that breed in plaque, the film of protein that collects on our teeth, and the acid corrodes the tooth enamel forming cavities in which yet more bacteria can thrive.

Dentists did little more than pull out rotten and painful teeth, of which there were many, but they eventually started to make false ones to fill the gaps. Some even used *human* teeth for this purpose and in 1781 a practitioner of Gerard Street, London, was offering to buy real teeth for £2 each, equivalent to something like £400 (approximately €600 or $700) in today's money, provided they were in good condition. (Such teeth were extracted from newly buried corpses by 'resurrectionsts' who dug up bodies to sell to surgeons for their anatomy lectures.) Other dentists fitted imitation teeth made of porcelain but these, first produced in 1774, were brittle and made an irritating squeaking noise when chewing. Across the Atlantic, George Washington had a set of dentures consisting of hippopotamus ivory into which were mounted teeth made from those of horses and donkeys.

The biggest supply of human teeth came as a result of scavengers plundering the bodies of the dead after major battles. The Battle of Waterloo, fought on 18 June 1815, was a particularly fertile hunting ground with around 50,000 casualties to pick from, and dead young men were more likely to have teeth in good condition. Thousands of teeth were extracted and used by dentists in the subsequent years. The dentures made from them were known as Waterloo teeth, and people were proud to be seen wearing them; some were even exported to the US. Later supplies came from the battlefields of the Crimean War of the 1850s, and the American Civil War of the 1860s, although by then it was more common to have dentures made of a stronger type of porcelain which had been invented in London in 1837.

Throughout history, people have tried to preserve their teeth by cleaning them. Back in the 4th century BC, Hippocrates suggested doing this with powdered marble, which is calcium carbonate, and indeed this chemical (as chalk) is still used as a mild abrasive in toothpaste. Toothbrushes were invented in China in 1498 and were known in Europe by the 1600s when they were on sale in Paris, as we learn from a letter sent to Sir Ralph Verney in 1649 asking him to buy one while he was visiting that city. They were used in conjunction with tooth powders. More convenient were the early toothpastes, which consisted of powdered chalk, soap, and sugar syrup. Toothpaste was first sold in ceramic pots and then it began to be sold in collapsible tubes that an American artist, John Rand, had invented in 1841 as a way of packaging oil paints.

Modern toothpastes have several components: the scouring agent is likely to be powdered calcium phosphate – or silica if the paste is a transparent gel – and the foaming agent will be sodium lauryl sulfate.[9)] This mild surfactant is added to help disperse the toothpaste in the mouth during the brushing action, thereby solubilising the plaque, and preventing any dislodged debris from re-depositing on the teeth. Toothpaste can have a curious side effect in some people, in that it

9) More about this in Chapter 3.

stimulates the bitter taste receptors on the tongue so that drinks like orange juice become very bitter when drunk immediately after cleaning your teeth. Another major ingredient is the humectant which keeps the toothpaste moist by retaining water and this accounts for about a third of the contents. The ones most commonly used for this purpose are glycerol or PEG (short for polyethylene glycol). The minor ingredients in a typical toothpaste are:

- an artificial sweetener such as saccharin, or the natural sweetener sorbitol;
- a thickening agent such as carboxymethyl cellulose, or sodium alginate, which is a carbohydrate extracted from seaweed;
- sodium benzoate, which prevents bacteria from breeding;
- fluoride, which strengthens the teeth against decay;
- flavouring, such as peppermint, or oil of wintergreen.

The aim of regular brushing is to keep teeth germ-free and looking good. Discolouration, however, comes slowly and cannot be removed by brushing alone.

Teeth have a transparent outer layer of enamel which is about 2 mm thick, and an inner layer of white dentine which surrounds the innermost pulp cavity where the nerves are located. Both the enamel and the dentine are calcium phosphate, a mineral that can exist in various forms, that of the enamel layer being hydroxyapatite, which is one of the hardest naturally occurring minerals. When this is exposed to fluoride it forms an even harder mineral: fluoroapatite. Unfortunately both this and hydroxyapatite are slightly porous and this is why teeth become stained by chemicals such as polyphenols and other dark coloured substances that are present in coffee, tea, red wine, bilberries, blueberries, and cigarette smoke. (Staining can also be caused by tetracycline antibiotics, especially when taken by children whose teeth are developing. This type of discoloration is permanent, which is why such antibiotics are now only prescribed for adults.)

The answer to discolouration is to bleach the stains using tooth whiteners. An earlier method of doing this was to use nitric acid which is an oxidising agent and capable of removing stains, but it also removed a layer of the enamel as well. Tooth whiteners now rely on →hydrogen peroxide as the bleaching agent and sales are in excess of $1.5 billion a year in the US alone, and the rest of world is catching up on the fashion for sparkling white teeth, as popularised by young television and movie stars. A dentist can bleach teeth very quickly by using a paste containing 35% hydrogen peroxide, sometimes in combination with a laser beam – supposedly to speed up the process, although the need for this is questionable – and stains will be gone within the hour. The paste is applied and rinsed off several times during the treatment, and it has to be carried out by a dentist because hydrogen peroxide of this strength can damage the lining of the mouth if not carefully applied. Dentists can also provide a treatment to be used at home. They make a mould of the teeth into which a strip of peroxide gel can be inserted and then pressed against the teeth. This 'nightguard' can even been worn while sleeping, thereby speeding up the whitening process.

For those who prefer a less expensive way of bleaching their teeth, there are products which contain a chemical that reacts with water to release hydrogen peroxide. This chemical is carbamide peroxide, also known as →urea peroxide,[10] and this is the ingredient in several over-the-counter whitening agents. Carbamide peroxide is made from urea and hydrogen peroxide. The former chemical is perfectly safe because it is a normal end-product of our body's metabolism and is one of the ways we dispose of unwanted nitrogenous material via our urine. In earlier times it was even prescribed by doctors because of its diuretic action, in other words it encouraged the body to remove excess water by stimulating the kidneys. When carbamide peroxide is in contact with water it releases its hydrogen peroxide which then gets to work, although a lot of this is lost to enzyme breakdown before it can do its job. This is why over-the-counter tooth whitening products need to be applied many times to achieve maximum effect. The most convenient way of whitening teeth is to stick a polyethylene strip bearing a peroxide gel to the teeth and leave it there for 30 minutes. If this procedure is repeated every day then after a couple of weeks the teeth will be noticeably whiter.

Those who engage in tooth whitening should be aware that it has a weakening effect on the enamel. This was reported at a meeting of the US Materials Research Society in Boston in November 2005 by Michelle Dickinson who works for Hysitron, an instrument maker based in Minneapolis. This company has developed a piece of equipment capable of measuring the hardness profile of teeth across the enamel and dentine

10) It also has trade names such as Exterol, Hyperol, Perhydrit, etc.
See also **urea** in Glossary.

boundary. Dickinson examined the effect on extracted human teeth of the carbamide peroxide solution of over-the-counter whiteners, and of the much stronger paste used by dentists. She subjected teeth to seven one-hour treatments of each kind of whitener and found that the former decreased tooth hardness by 22% and the latter by a worrying 82%. The result would be an increased sensitivity of the teeth to hot and cold. These findings have yet to be confirmed, but they are a warning that tooth whitening should not be undertaken too often.

There are ways of obtaining an instant gleaming smile, such as having your teeth covered with resin or porcelain veneers. The cost of these can be prohibitive, with the porcelain ones being twice as expensive as the plastic ones. Over time, these too can become stained.

So what might we expect in the future? Ideally we should try and make tooth enamel impervious to stains. One product that might soon become available is a chewing gum which contains both calcium and phosphate to help teeth to repair cavities by boosting the natural level of these components in our saliva, part of whose job is to repair teeth – see box. However, the best agent for strengthening teeth is still fluoride.

The Chemistry of Saliva

The average mouth excretes half a litre (500 ml) of saliva a day. This does not just lubricate the mouth and aid digestion, it contains substances that benefit the teeth, among which are calcium ions (120 mg per litre of saliva) and phosphate (14,000 mg per litre of saliva). The **pH** of saliva is effectively neutral, being 6.8, and this is ideal for the tooth enamel to reabsorb calcium and phosphate to repair itself. If the pH falls below 5.5, then the reverse happens and there is some loss of these components from the teeth.

What saliva also supports are the millions of bacteria living in our mouth, of which more than 300 different types have been identified. While this sounds alarming, it is not too worrying because most of them are harmless and some even appear to protect the mouth. Anthony van Leeuwenhoek first demonstrated the presence of bacteria on teeth as long ago as 1683, using the microscope that he had invented, and he described what he had found as follows:

> *"...in the said matter [plaque] there were many very little living animalcules. There are more animals living in the scum on the teeth in a man's mouth than there are men in the whole kingdom – especially in those who never clean their teeth, whereby such a stench comes from the mouth of many of them that you can scarce bear to talk to them..."*

That was more than 300 years ago and yet oral hygiene is still a problem. The answer is to clean the teeth and gums by brushing and rinsing, and to stimulate saliva by chewing gum.

It has been known for more than 200 years that teeth naturally contain some fluoride, and we now know that this element strengthens tooth enamel by forming fluoroapatite, which resists the effect of the acids formed by oral bacteria. The fluoride also inhibits bacteria from multiplying. For this reason fluoride is added to toothpastes and public water supplies. The average diet provides up to 3 mg of fluoride a day, depending on how much fluoride-containing food a person eats, such as chicken, sardines, mackerel, salmon, eggs, and potatoes, and on how much tea they drink, with a cup of this beverage providing 0.4 mg. Using sea salt to flavour their food will add more fluoride because the sea contains 1 mg per litre. Procter & Gamble introduced the first fluoride toothpaste, Crest, in 1955 and it contained tin fluoride (SnF_2) as the protective agent, this having been discovered to be the form which worked best, thanks to research by Joseph Muhler at the University of Indiana in the 1940s. This fluoride was eventually replaced by sodium monofluorophosphate (Na_2PO_3F) which works even better.

Another protective element is strontium. The regular version of Sensodyne toothpaste contains 10% strontium chloride, which has the advantage of helping to build up the enamel, especially round the gum line. As gums begin to shrink with age they can expose the dentine layer making a tooth very sensitive to the brushing action of the toothbrush, as well as to heat, cold, and acids. There are lots of tiny tubes (tubules) in the dentine and these contain fluid which reacts to changes in temperature or pressure and transmits a signal to the tooth nerve which then registers intense pain. Strontium helps to block them.

Soon there may be another ingredient in toothpaste: hydroxyapatite nanoparticles[11]. Not only is this the same chemical as the tooth enamel but the particles are small enough to get into the pores and seal them, and it is pleasingly white. Ralf Nörenberg of the chemical giant BASF reported the new form of calcium phosphate in 2003. More recently another group of researchers, led by Kazue Yamagishi, and based at the FAP Dental Institute in Tokyo, have developed a synthetic enamel based on the same nanoparticles. They observed that when this was applied to a tooth, along with hydrogen peroxide, then new crystals grew inside the tooth's tubules and within 15 minutes these

11) A million nanoparticles measure around 1 millimetre.

crystals had bonded to the natural tooth enamel. It is more than likely that one day there will be toothpaste based on hydroxyapatite nanoparticles.

Nailed

When people meet for the first time they surreptitiously examine each other: clothes, face, hair, and teeth come under scrutiny and we have already discussed how to improve some of these. They may also glance at each other's hands and then they may judge a person by the condition of their fingernails. Clean, well-cut nails suggest self-confidence and careful attention to detail. Other conditions of the nails may send out less-flattering messages. A man with long fingernails may be thought slightly odd and probably without a partner, while someone with bitten fingernails may be seen as nervous and stressed, and dirty fingernails will speak of slovenliness and lack of personal hygienic. Men take little interest in their fingernails apart from cutting them regularly and keeping them clean. Women on the other hand have turned fingernails into a minor art form.

Many women simply want carefully shaped, well-manicured nails, and many of those who paint their nails use unobtrusive shades. On the other hand there are those who enjoy showing off their nails with brightly coloured varnishes, and may even enhance them with artificial extensions on which are painted intricate designs. Media celebrities and the wives of high profile sportsmen seem particularly attracted to wearing them as glamour accessories. All over the UK and the USA there are nail salons and nail booths, manned by manicurists who now prefer to be called nail technicians. Even the ancient market town where I live has two nail salons. Its clients may not appreciate the role of the chemist in meeting their needs, but a lot of research has gone into nail varnish and nail extensions.

Nail varnish should offer a wide range of colours and textures, adhere fast to the nail, and not chip. It should also be water-resistant, but easy to remove with a non-hazardous solvent that does not damage the nail, the skin, or the environment. All these conditions have been met with a blend of colours, polymers, plasticizers, and solvents. Nail varnish contains a pigment or dye, plus nitrocellulose to provide a gloss, and butyl stearate plasticizer to keep the varnish flexible when it has dried and to ensure that it doesn't chip. Toluene sulfonamide formaldehyde (TSF) may be added to increase the strength of the final film because this is strong and durable. There are some nail varnishes that do not include nitrocellulose – manufacturers would like to remove this because it is highly flammable and explosive in bulk – and they use methacrylate polymers instead. The various ingredients of a nail varnish are suspended in a mixture of solvents such as acetone, toluene, isopropanol, and pentyl acetate, designed to give a varnish that runs easily when applied, but dries quickly. Isopropanol is used specifically to hold particles such as glitter in suspension. Before applying varnish to a nail it might be necessary to remove a cuticle and there are solutions for doing this, and these consist of potassium hydroxide (KOH) in a solvent mixture consisting of 12% glycerol and 88% water. Eventually the wearer will want to remove old nail varnish and then it is necessary to wipe it off with a solvent like acetone or ethyl acetate, and this will contain small amounts of things like glycerol and lanolin, which are there to rehydrate and replenish the natural oil of the nail and the surrounding skin.

Polymers for artificial nails first appeared in the 1970s. Before then there were porcelain ones but these were brittle. The new polymer nails were much better and they could be trimmed with scissors and smoothly shaped with a nail file. Nail extensions are the more technical side of nail culture and they have been made from various plastics but are usually either polyacrylate or the copolymer ABS, short for poly(acrylonitrile-butadiene-styrene), which has the stiffness and flexibility very similar to those of natural fingernails.

Ideally an artificial nail should cover about half the real nail whose surface has to be roughened slightly before an adhesive is applied and the artificial nail stuck on. How far the extension protrudes beyond the end of the finger is up to the client but this will increase as the natural nail grows, which is why trim-ability and file-ability are essential. Artificial nail tips can be glued to the existing nail with an adhesive like rosin, which is the sticky residue left when tree oils have been distilled to remove their volatile oils. Alternatively, the artificial nail may be bonded to the real nail with methyl cyanoacrylate (popularly called superglue). This is a chemical that remains stable until it absorbs a little water from the air, whereupon it immediately starts to polymerise to form a tough resin that will stick two surfaces together permanently. The glue can work within 10 seconds. Equally good are ethyl cyanoacrylate and butyl cyanoacrylate. The vapour from superglue is unpleasant to breathe and if there is more than 2 ppm in the air it is intensely irritating, which is why the less volatile ethyl form is sometimes used. However, though ethyl cyanoacrylate is a more preferable adhesive, it can cause an adverse reaction in some people with the result that the natural nail withers and the ends of the fingers develop eczema. Three such cases came to light in the Department of Medical Sciences at the University of Arkansas in 1998. The US National Institute for Occupational Safety and Health (NIOSH) has issued guidelines for nail technicians, the main suggestion being that they should sit at specially ventilated tables that extract any vapours given off by the chemicals being used.

Having secured the nail extension, and smoothed over the joint it makes with the real nail, the next job is to cover the whole with a film that inconspicuously unites the two parts. This is done by applying a paste made from powdered methacrylate polymer which is smoothed on to the nail where it will harden by absorbing oxygen from the air. Sometimes a little benzoyl peroxide can be added to speed up the process, sometimes it is hardened by exposure to UV light, and there are even some films that harden under ordinary light. Several layers of gel are applied until the desired smoothness from the base of the nail to the tip has been achieved. Finally the nail can be painted and decorated, sometimes to stunning effect with tiny diamonds.

Of course nothing can stop such nails from moving up the finger and eventually they have to be removed. The solvent for doing this is acetonitrile[12)] which must be used carefully because it can be absorbed through the skin, which is why it is only available at nail salons.

Of course there are dangers in having synthetic nails. One worry that preoccupied American chemists a few years ago, and led to papers in the *Journal of Chemical Education,* was the possibility that they might be dangerously flammable. If young chemists had to use a naked flame, such as that of a Bunsen burner whose temperature is around 500 °C, then serious burns might be caused if they wore plastic nail extensions. The paper reported that an artificial nail would ignite in less than a second. Even when they come into contact with the flame of a candle on a birthday cake, which has a much lower temperature, they can still ignite in just over a second. Once ignited, they

12) Chemical formula CH_3CN.

start to curl and drop molten balls of plastic. If the burning finger is shaken, which is the natural reaction to what is happening, it only makes things worse. The upshot was a reminder that students in chemistry labs should not wear such fingernail extensions. No doubt there are other situations involving a naked flame when these nails should not be worn, such as when cooking over a gas-fired stove or while using matches, and nail technicians can no doubt recount stories of clients who have suffered such accidents.

Bacteria pose a bigger threat than fire. Natural fingernails, or rather the dirt beneath them, account for 80% of the microbes on the hands. Some of the bacteria, yeasts, and fungi that live there can be dangerous. Artificial fingernails and nail extensions are even more dangerous, especially if they are worn by health care workers and nurses. Not only do they harbour more bacteria, they may also puncture latex gloves. For this reason it has been made illegal in the US for those in the caring professions to wear artificial nails, and this means nurses, doctors, and therapists. According to guidelines issued by the US Centers for Disease Control in 2000, these groups of workers should always have nails shorter than the tips of their fingers and should be well scrubbed.

Outbreaks of disease have been caused by those with artificial fingernails. In 2004 there was an outbreak of *Klebsiella pneumoniae* among premature babies in a US intensive care unit, caused by bacteria from a nurse's artificial nails. A few years previously it was *Pseudomonas aeruginosa* that threatened several newborn babies in a New York hospital and this was traced to the same cause. In Canada three patients who had had surgery on their spinal cord developed *Candida* infections of the spinal disks and this was traced to an operating theatre technician who had artificial nails. An intensive care unit in Oklahoma City saw 16 patients die as a result of contracting *Pseudomonas aeruginosa* from two nurses who had artificial nails. Thankfully such outbreaks are now extremely rare.

For many women, painted nails and artificial nails are harmless affectations that give pleasure and there is no need to worry about them. While their wearers may not acknowledge their debt to chemistry, they nevertheless are gentle reminders of the way chemistry can help us feel better about ourselves. Sadly many still see 'chemicals' as an ever-present danger and worry unnecessarily. The following item may serve to assure them that the risks to health are negligible.

Issue: Are Natural Ingredients for Cosmetics Better than Those Made by the Chemical Industry?

It sometimes says on the packaging of cosmetics that the contents are 'organic', 'pure', and 'natural', suggesting by these and similar terms that the products are better than those made from synthetic ingredients. Nothing could be further from the truth, if only for the simple reason that natural materials are not quality controlled and are often impure while those from the chemical industry are quality controlled and pure.

Some natural impurities can cause an allergic reaction and this is true of traces of enzymes, which are large molecules that living things produce in order to carry out essential chemical reactions inside living cells. A person's immune system may perceive unfamilar enzymes as a threat and will attack them, the result being the uncomfortable symptoms of an itchy rash, hives, inflammation, runny nose, headache – or worse. It is possible to become allergic to a synthetic chemical, as we saw with the PPD used in hair dyes, but such allergies are exceedingly rare.

Another threat which comes with natural products is microbes. The ingredients in cosmetics, such as water, oils, carbohydrates, minerals, and proteins, make them an ideal medium in which bacteria can multiply, as some purchasers of purely natural beauty products have discovered to their cost, not least of which is the disagreeable odour they emit which indicates they are going off. Even such natural products must now contain anti-bacterial agents and these must be proven to work, which is why they are generally synthetic chemicals. The ones most used and most effective at killing bacterial are the parabens. These are simple molecules that are modified versions of a naturally occurring fruit acid,[13] yet rather perversely there are those who campaign against them on the grounds that they are 'chemicals', meaning they are products of the chemical industry, and that of course is where they are made.

In fact Nature is much more prolific than chemists at making chemicals, and among them are some that really do have healing benefits, not that Nature designed them with this in mind. Traditional cures may be based on a plant or marine extract and it then becomes a challenge for chemists in the pharmaceutical industries to make exactly the same molecule in the laboratory. If it is truly beneficial, and without harmful side effects, then a way may be found to manufacture the material on a larger scale. Even if tests show the natural chemical has harmful side effects, it may be possible to change the molecule in a way that keeps the active centre but eliminates the dangerous part or replaces it with something that is much safer.

Examine the list of contents of most cosmetics and you will see a bewildering array of chemical names, but you can be sure that these have all been manufactured to agreed standards of purity and have been tested to ensure they are safe to use. By all means buy the cosmetics that appear to be 'nature resourced' but don't imagine that this somehow confers an added benefit – it doesn't. And don't fool yourself into thinking that because it is 'free of chemicals' that you are somehow protecting yourself – you aren't.

13) This is para-benzoic acid, more correctly called 4-hydroxybenzoic acid, found in strawberries and grapes.

2

Better Looking (II) and Better Living (I): Medical Advances

A small arrow printed before a word in the main text indicates that there is more information on that topic in the Glossary.

News from the Future

Global Times News, 21 March 2025

Ms World in €5M Product-placement Row

The promoters of this year's Ms World beauty contest were dismayed when the winning contestant Heidi Hussein (Ms Germany) thrust a tube of the skin cream AxAcne at cameras filming the show live on Saturday. Her gesture, seen by more than 3 billion people worldwide, brought the makers of the anti-acne cream publicity that some estimate would have cost more than €5M if purchased as product-placement advertising. Neither Heidi Hussein nor Ms World Productions would reveal how much they had been paid for what one commentator called 'a corrupt publicity stunt.'

BrazilChem, makers of AxAcne, said that Ms Hussein's agent had contacted them when she became one of the 10 short-listed contestants, saying that as a teenager she has suffered badly from acne and that she had used their product successfully and felt she 'owed it all' to AxAcne.

"It is our intention to sponsor many of the events Ms Hussein will be taking part in during the coming year" said company spokesperson Dr Tracey Schmidt. 'We see nothing wrong in this. As you know, millions of young people suffered from acne down the ages leaving them with permanently scarred complexions, but this is a thing of the past thanks to gamma-hydroxy-pentanoic chloroester which was discovered at the Brasilia Dermatological Research Center ten years ago.

Page 2: Pictures of Ms Germany before and after acne treatment.

Page 3: AxAcne may cause skin cancer later in life warns top plastic surgeon.

Better Looking (II): In the previous chapter we looked at ways chemistry can help people appear more attractive, but there are conditions where cosmetics are of little use because the problem is much deeper, which brings us to the first topic of this chapter: skin disease. Skin diseases are not life-threatening but they can and do cause life-long resentment. Sufferers wonder: why me? Some may even be driven to

Better Looking, Better Living, Better Loving. John Emsley

ISBN 978-3-527-31863-6

hiding themselves away and some are even driven further, to ending their lives. What can chemistry do to rescue them from their lonely islands?

Later in the chapter we will look at the relief that can be brought to people who suffer the pains of old age, we will discover that carbohydrates are not just food components but have a role to play in protecting us, and we will consider the molecules that are used as modern anaesthetics. Finally we will discuss the safest treatment of all: homeopathy. Can it really provide a cure? The answer may surprise you.

Skin Diseases: Acne, Eczema, and Psoriasis

Nothing undermines a person's confidence like skin disease, especially when it is visible on the face. Acne can strike at a time when a young person is socially most vulnerable, but other skin diseases can be equally disturbing, such as eczema, which leads to severe itching, scarring, and loss of sleep, and psoriasis which, while not generally visible, can restrict a person's activities, especially if this means exposing parts of the body that are badly affected. There is also vitiligo, a disease in which areas of the skin are completely white and while this is rare in Western societies, it too affects the quality of life of those who suffer from it, especially in India. Vitiligo is mentioned in the sacred Hindu text the *Atharva Veda*, written 3,400 years ago, and it is popularly referred to as 'white leprosy'. It used to be treated with the juice of the bavachee plant[14)], and even in those days the doctors knew that after a patient had had the plant juice applied to the white patches on their skin, they had then to sit in the sun for it to take effect. Today they use methoxsalen, which also needs to be exposed to strong light, and this is based on the natural chemical which plants like bavachee produce.

The skin is not just the protective packaging of the body; it is a large and complex organ. Indeed it is the largest human organ, with an area of 2 square metres. It can be thin, as eyelids, or thick, as on the soles of the feet. It constantly grows from the inside and is constantly lost from the outside, and we shed around 2 g of dead skin cells per day as dust. The outer layer of the skin is known as the epidermis, below that

14) The juice of the bishop's flower plant was also effective.

is the dermis. The epidermis has a waterproof top layer, which is the body's first line of defence, and a lower layer (the basal layer) where new skin cells form. This basal layer is geared up for wound healing, a process that involves white blood cells, and it has chemicals known as growth factors which promote cell division. There are also melanocytes and these are there to shield the body against damaging ultraviolet rays by forming the dark-coloured chemical melanin. The deeper dermis provides extra protection with collagen, a tough fibrous protein that is both strong and flexible. The dermis is also equipped with sensors to detect heat, cold, vibration, pressure, and pain.

As we shall see in Chapter 3, the skin is well provided with eccrine and apocrine sweat glands, and there are also sebaceous glands which are closely associated with the hair follicles and provide an oily secretion known as sebum. When this becomes oxidised by the oxygen of the atmosphere it turns black and then we have a blackhead. This disfiguration is relatively innocuous and easily dealt with. Acne, eczema, and psoriasis, however, are much more serious and they involve the structure of the skin. Acne affects mainly the hair follicles and sebaceous glands; eczema is an immune condition that lifts layers of the epidermis causing inflammation and itching; psoriasis is abnormal cell division. Although these three skin diseases are very different in their causes, the medicines used to treat may be the same.

A traditional treatment for any skin condition was to rub on some soothing ointment and lanolin, the chemical that makes the wool of sheep waterproof, was often used and provided some relief by replacing or supplementing the skin's own natural oils. A better treatment came indirectly as a result of the introduction of gas lighting in the 1800s. The gasworks where coal was heated to release gas also produced large quantities of various chemicals, including coal tar and it was this which contained an agent that really was beneficial in the treatment of skin diseases. Indeed coal tar extracts are still used to control psoriasis. Another remedy for psoriasis was devised by Dr Balmanno Squire in 1876 and this was goa powder which was extracted from the Brazilian araroba tree. Chemical analysis later showed that its active ingredient was a derivative of the chemical anthracene[15] which is also a component of coal tar.

15) Anthracene consist of three benzene rings fused together, and has the chemical formula $C_{14}H_{10}$.

A major advance in treating skin diseases came in the 1930s with the investigation of the hormone chemicals produced by organs of the body, such as the adrenal gland. Studies showed that such chemicals could have profound effects on the skin and the steroid drugs based on them still have a role in treating eczema and psoriasis. In the 1980s, chemicals related to vitamin D were found to be beneficial in controlling skin disease and the drugs calcipotriol, calcitriol and tacalcitol were introduced. In the 1990s research moved from control to cure because there was now a better understanding of what was happening to the skin in these three diseases. Research has continued into the 2000s and currently dozens of new pharmaceutical drugs are proceeding through the various stages of →drug trials, and with a bit of luck some will be available for doctors to prescribe in a few years time.

Acne

This is the most common skin disorder, affecting mainly teenagers, and 9 out of 10 people experience some degree of acne in adolescence. It manifests itself as pimples, whiteheads, pustules, and even cysts, and in serious cases the skin can be left scarred for life. Acne generally clears up by the time a person reaches 20, but for one person in 200 it persists into adulthood and can be very disfiguring.

Acne is caused by changes in the skin and the most important of these is an increase in sebum production, due to an increase in sex hormones at puberty. Excess sebum can lead to an excessive growth of bacteria and particularly *Propionibacterium acnes,* which feeds upon it. These microbes produce irritant chemicals which seep into the dermis where they cause the sebaceous glands to swell and even burst, whereupon white blood cells rush in to fight invading organisms and so we have the pus-filled 'zits' that so characterise acne. The face has most sebaceous glands, which is why acne generally manifests itself there, although it can also affect the neck, shoulders, upper back, and chest.

There are other forms of acne apart from teenage acne. There is the industrial disease called chloracne caused by exposure to chlorinated oils, but this is now extremely rare. Certain steroid drugs can also cause acne because these stimulate the production of sex hormones. Another type of acne is caused by sensitivity to certain cosmetics, and there is contact acne which comes as a result of repeatedly touching a sensitising surface, as in the condition known as violinist's chin.

Acne treatments can attack the disease from more than one direction. They reduce the amount of sebum being secreted and so prevent blockage of the ducts, they kill the *Propionibacterium acnes* bacteria that are causing the disease, and they act to reduce inflammation. The simplest way to control acne is to dab the skin with one of the proprietary products, like Clearasil.[16] These over-the-counter preparations contain active ingredients like triclosan, benzoyl peroxide, or salicylic acid which have an anti-bacterial action, plus something to dissolve the plug of infected sebum which is blocking the pores. Nicotinamide may also be an ingredient because this will reduce inflammation. If this type of treatment proves ineffective then something stronger can be prescribed by a doctor.

When acne is severe it may require medicines that have to be taken by mouth and these may be stronger antibiotics combined with hormone therapy. Tetracyclines and erythromycin are typically used and they have to be taken for many weeks, but there are signs that bacteria are becoming resistant to these antibiotics and if there is no improvement in the acne within three months then this type of treatment has to be discontinued, or a much stronger antibiotic tried. Hormone therapy is available for young women with severe acne and this can block the sebum-stimulating effect of natural hormones – as well as having a contraceptive effect.

→Isotretinoin was introduced in 1982 and had a dramatic effect on some forms of acne by blocking the formation of sebum, which it does after about a month. Isotretinoin is normally only used under the supervision of a hospital dermatologist because it is so powerful and it has several side effects such as dry skin, cracked lips, sore eyes, and even nose bleeds. It raises the level of fats and cholesterol in the blood, which some researchers have suggested could have long-term effects.[17] Despite these drawbacks, it has helped those whose lives were being severely disrupted by acne and it is widely prescribed, particularly if there has been a relapse after a course of antibiotics, or where there is clearly emotional distress, or where permanent scarring is likely to occur. Isotretinoin must not be used if there is any chance of

16) There are various preparations marketed under this brand name, which was launched in the US in 1959.

17) Alternatively isotretinoin can be applied as a skin cream for direct application to affected areas.

a women becoming pregnant because it can cause birth defects, so it cannot be prescribed if there is any chance of this happening, unless she can convince a doctor that both she and her partner are using contraceptive methods. Even when a woman has successfully completed a course of isotretinoin she should avoid becoming pregnant for at least a month. Isotretinoin is initially prescribed at a rate of 25 mg per day for up to 4 weeks and is then increased to 50 mg for a further 8 weeks.

How successful is isotretinoin? For 40% of patients it cures acne permanently, while for 20% it requires a further course of treatment before the acne clears up, but most who use it experience some benefit. It has to be taken with a fat-containing food because the molecule is fat soluble and this aids its absorption by the body. Isotretinoin received some bad publicity in the media when the son of a US congressman committed suicide while taking the drug. However, no link has been established between isotretinoin treatment and an increased incidence of suicide compared to those undergoing other treatments for acne. Good as isotretinoin is, it is still far from a guaranteed remedy. Sadly, acne will continue to blight the lives of some people, but at least today there is a reasonable chance that even severe acne can be cured.

Eczema

The name of this disease comes from the Greek word meaning 'to boil over' which is how it appears in its most acute form when there is severe blistering of the skin, as if a person has been scalded with boiling water. Normally eczema is less severe than this and appears as scaly skin which causes intense itching. This results in scratching, especially by young children, which leads to secondary skin damage through infections. Often eczema has no clear cause and if it is going to develop it will do so before the age of one and as many as ten per cent of babies are affected. It will often start as a mild rash on the face but by the age of two it will be most troublesome on the wrists, in front of the elbows, and behind the knees. Eczema can mar the childhood years, but it will generally disappear soon after puberty, although it may recur during later life.

One form of eczema, known as seborrhoeic eczema, affects areas around the sebaceous glands of the face, scalp and chest but it is relatively mild and only becomes a nuisance if it gets infected with bacteria, or yeast microbes such as *Candida*. This type of eczema can be kept under control with medicated soaps and shampoos. The other common type of eczema is referred to as atopic eczema and while there is no obvious irritant causing it, suspicion generally falls on things such as milk, house dust mites, wool, pets, cigarette smoke, and the fragrances used in toiletries. The nickel component of stainless steel, as used in garment fasteners, ear-rings, and watch straps will cause eczema in some people. Once this has been recognised then evasive action can be taken. Another form of the disease is contact dermatitis in which blood vessels in the skin are engorged, the surrounding tissue is spongy due to excess fluid, and there are clusters of inflamed cells. This type of eczema is triggered by an irritant which damages the skin and then penetrates deeper, where it triggers →T-cells to proliferate and release all kinds of molecules that cause further inflammation. There are forms of eczema that afflict the elderly, often appearing as a side effect of other illnesses. Eczema can be complicated by becoming infected by the bacteria *Staphylococcus* or *Streptococcus*, or by the virus *Herpes simplex*, or by fungi.

Eczema may be so mild that no treatment is required beyond avoiding known irritants and applying a soothing and protective ointment which keeps the skin moist and so reduces inflammation. There are many such over-the-counter remedies based on liquid paraffin and soft white paraffin, both of which are hydrocarbons obtained from refined petroleum and which contain nothing that can irritate the skin. Liquid paraffin is also called mineral oil and is available under a variety of trade names, such as Nujol, while soft white paraffin is better known as Vaseline. Emulsified mixtures of liquid paraffin and soft white paraffin make excellent skin creams.

Steroid creams will often clear the condition and hydrocortisone can be prescribed under a variety of trade names such as Alphaderm. More powerful are the creams based on betamethasone esters, of which Betnovate is the best known and has been used for more than 40 years. If this does not work then the doctor can prescribe even stronger drugs based on clobetasol propionate, such as Dermovate, or similar active agents. There are several such creams now available and it is rare for eczema not to respond to one of them and be cured – and kept at bay. If the eczema is infected then various antibiotics, antivirals, and anti-fungals can be given to clear up the infection. Very itchy eczema can be controlled with antihistamines. In extreme cases of eczema a doctor can prescribe powerful drugs, such as cyclosporin and azathioprine, designed to suppress an over-active immune system.

Significant progress in treating eczema came with the introduction of the drug tacrolimus which had been used for many years to prevent tissue rejection in transplant patients. It blocks the T-cell action and in those with severe eczema it also results in a marked improvement in the condition. Another drug, pimecrolimus, began to be used in 2003 and has been specially designed for applying directly to the skin, even of young children, and it too works by blocking the T-cells and stopping the inflammation, redness and itchiness.

Psoriasis

This appears as areas of reddened and flaky skin and can be a lifelong condition although it tends to flare up in the teens and twenties and then again in old age. About 1 person in 50 suffers from psoriasis at some time in their life. Why it occurs is still not understood. Psoriasis is caused by over-reactive skin cells in the lower layer of the epidermis dividing 20 times faster than normal. A normal skin cell takes around 4 weeks to mature and reach the surface of the skin, there to be shed. Psoriatic cells go through this process in only two days and they accumulate at the surface as a layer of dead skin. Skin affected by psoriasis has a thickened epidermis with an excessive growth of blood vessels, and there are clusters of immune cells. Plaque psoriasis is the most common and occurs on the knees, elbows, lower back, and scalp.

Early treatments for psoriasis were based on coal tar products and these continue to be used with some success and we now know that they worked by blocking the synthesis of DNA in skin cells, thereby stopping them from proliferating. Coal tar creams and ointments prevented skin from cracking and reduced inflammation. Another treatment of long standing is dithranol which is used mainly in hospitals and is painted on the skin, but it is not popular with patients because of its intense purple colour which stains skin and clothes. It must not be left on the affected area for more than 30 minutes but it does reduce the severity of the disease. Several other ointments have been used to treat psoriasis with moderate degrees of success. Steroids such as hydrocortisone may work, but they can have long-term side effects and have now been superseded by other medicaments. Tazarotene is another skin cream which will control mild plaque psoriasis.

Medicines related to vitamin D are also effective but an unwanted side effect of the early ones was to cause the body to retain too much calcium, which is detrimental. More than 1500 different compounds have been synthesised to try and find one that would act on the psoriasis without raising the level of calcium in the blood. Calcipotriol cream was the one that performed best and is now widely used. Another chemical, etretinate, was introduced in 1975 and found to be just as effective. It was eventually superseded by acitretin when it was discovered that the body converted etretinate to acitretin and it was the latter which was the active agent. Acitretin is now the prescribed drug and it is very effective with three quarters of those treated reporting noticeable improvement and for about a third of patients their psoriasis cleared up completely. Psoriasis can also be treated with successive oral doses of methoxsalen followed by UV irradiation of the affected area.

Psychological treatments may even be effective because the condition seems to be caused by stress. Deep breathing, meditation, and even listening to relaxation CDs can help develop a positive attitude that can speed recovery.

The release of chemical signals called cytokines appears to be the root cause of psoriasis and there are many kinds of cytokines. Some trigger cell multiplication, some cause abnormal cell development, and some cause inflammation. Only by separately identifying them was it possible to understand what cytokines were doing and then medicines could be devised to counteract their actions. Severe psoriasis can be attacked with immunosuppressants but these are powerful and have to be used carefully because of side effects. Alefacept and etanercept are a new class of drugs for controlling the immune system. They consist of a combination of two human proteins and were introduced in 2003. Alefacept blocks receptors on overactive T-cells, while etanercept deactivates TNF (short for tumour necrosis factor) which is known to stimulate psoriasis. Research on TNF was carried out in the 1990s by Ravinder Maini and Marc Feldman of the Kennedy Institute for Rheumatology, London, and they shared the 2000 Crafoord Prize of $500,000 (plus gold medals) awarded by the Swedish Academy of Sciences for research in areas not covered by Nobel prizes.

Many new drugs are currently being tested for treating acne, eczema, and psoriasis. For eczema there is likely to be a treatment based on desensitising the skin through an injection of enzymes and allergens. In the case of psoriasis a new approach is being followed in which proliferation of blood vessels in the affected area of skin can be stopped by inhibiting their growth factors. There are even some natural products that have been discovered and are being tested that can act this way. At the time of writing more than 50 drugs are in various stages of development and testing for these conditions. It may well be that the good news from the future, as in the item at the start of this chapter, will come sooner than 2025.

An Agile or Fragile Old Age?

The word arthritis simply means inflammation of the joints and the disease has been recognised as an affliction of the human condition since early times. The Hindus of India wrote about it as long ago as 1000 BC, and Hippocrates in ancient Greece believed it to be a symptom of poisoning which could be relieved by blood letting. A hundred years ago doctors thought it was caused by an infection and one way to treat it was to remove sources of infection from the body, such as teeth, tonsils, and appendix. Today we regard arthritis as an autoimmune disease in which the body's defence mechanism has turned upon itself.

During the first half of the last century the usual treatment for arthritis was aspirin, and this worked in that it relieved the pain and reduced the inflammation around affected joints. Occasionally it was found that drugs designed to treat other diseases also relieved the symptoms of arthritis, and this was how gold came to be prescribed. Bacteriologists in 1890 had observed that gold cyanide prevented the tuberculosis bacterium from multiplying, but it was not until the 1920s that the chemical gold sodium thiomalate (better known as Myocrisin) was tried as a cure for tuberculosis, against which it turned out to be useless. However, the investigating doctor, Jacques Forestier, noted that it relieved rheumatoid arthritis and in 1935 he reported on a six-year study that showed Myocrisin did in fact slow down the course of the disease. Myocrisin had to be given by deep intramuscular injection starting with a test dose of 50 mg, and if this was accepted by the body then a course of weekly injections were given until there was evidence of improvement in the condition. For 1 person in 20 Myocrisin provoked a serious reaction and for a few people the injections proved fatal.

Another gold drug, auranofin (trade name Ridaura), was introduced in the mid-1980s after studies on 4000 patients proved positive. This can be taken orally but if there is no sign of it working within 6 months then it has to be discontinued. To get the body to absorb the gold its atoms are attached to sulfur atoms which are attached to a carbohydrate. In this form it can pass through the stomach wall and into the blood stream, and it eventually ends up as gold cyanide which is the active form of the drug and is known to inhibit the immune response. (The cyanide comes from that which is already naturally present in the body, albeit only in tiny amounts.) Gold cyanide also binds to albumin in the blood and this will benefit people with arthritis because albumin is abnormally high in those with the disease. Gold therapy is now part of the medical armoury against rheumatoid arthritis, and treatment with it even merits a special name, chrysotherapy, from *chrysos* the Greek word for gold. It is prescribed when other drugs are failing to give relief. Gold can take up to ten weeks to work, but it goes on working even when treatment ceases, with some patients reporting improvement continuing for as long as a year after they stopped taking the drug. Even though a patient's symptoms are alleviated, the treatment has to stop after a few years because of a build-up of gold deposits in the body and side effects such as diarrhoea and skin rashes. Gold is not without its risks and gold-based drugs produce serious side effects in about 30% of those who use them, the main side-effect being an itchy skin rash which can be severe enough for the therapy to be discontinued.

There are two kinds of arthritis: osteoarthritis, which develops slowly and afflicts the elderly, and rheumatoid arthritis, which can come on quite quickly and generally affects those aged between 35 and 45. For these unlucky individuals onset is sudden in 10% of cases, while for another 20% it takes several weeks to develop, and for the remaining 70% it comes on slowly. Rheumatoid arthritis affects around 1% of the population and is thought to reduce the life-span of those affected by around 5 years. Osteoarthritis is more common with around 3% of the population affected. About one in seven of those suffering from arthritis require medication, but most people can live with the disease and rely on exercise, physiotherapy, or local warmth to relieve symptoms, and there are other treatments which help such as infra-red heat, ultra-sound, and hydrotherapy in a warm pool.

Arthritis develops in the membrane that surrounds the joints of the fingers, wrists, elbows, shoulders, feet, ankles, knees and hips. The membrane becomes enlarged and its cells begin to attack the cushion of cartilage between the bones and even to corrode the ends of the joint bones themselves. The result is painful inflammation. Cartilage is easy to damage but difficult to repair because it has no blood supply so cannot heal itself. Arthritis has symptoms in common with other diseases but it can be identified by following the guidelines issued by the American Rheumatism Association. The early warning sign is joint stiffness in a morning that lasts for an hour or so after getting up. More indicative are swellings around at least three joints and particularly the joint of the middle finger, and the fact that the same joints are affected on both the left and right sides of the body. There are medical tests that can help confirm arthritis, such as the presence of certain antibodies in the blood. These are found in 85% of those with the disease, although 5% of perfectly fit people also have them. X-rays of the affected joints will reveal definite changes.

There is some evidence that arthritis tends to run in families, suggesting a genetic factor is involved and this has been associated with a gene called HLA-DR4, but this occurs in a quarter of the population, the vast majority of whom never suffer from the disease. It may be that a virus triggers the disease, or even a bacterium, but none has been identified. Whatever the cause, the result is that fluid swells the joint tissue and this provokes a response from the body in the form of natural chemicals known as prostaglandins. Joints become inflamed when the immune system becomes activated. Normally the immune system is geared to fight an invading microbe from without, or rogue cells like cancer from within, and when the danger is over the immune system returns to its normal quiescent state. In arthritis the immune system continues to struggle against what it perceives to be a serious danger, meanwhile only making matters worse.

Even if we cannot yet cure arthritis, there are things we can do to alleviate its symptoms. Simple painkillers such as paracetamol, ibuprofen and codeine can be taken, and the combination of ibuprofen with a little codeine is particularly effective, a tablet will give relief for up to 6 hours. When such painkillers are no longer strong enough, then a person should seek medical help. After arthritis has been diagnosed by a doctor, treatment generally begins with a stronger non-steroid anti-inflammatory drug (NSAID) and various ones will be prescribed to reduce the level of prostaglandins. Some prefer to use NSAID creams that can be rubbed on to relieve painful joints, but these only work in the short term and after one month they appear to provide no relief at all. That was the conclusion of Weiya Zhang of the University of Nottingham, England, who carried out a survey on such creams in 2004. Which drug the doctor chooses, and continues to prescribe, will depend on the patient's response to it and whether there are unwanted side effects, of which the most common are ulcers, bleeding, and even perforation of the stomach wall. In the UK alone around 12,000 are admitted to hospital every year to treat these side effects. Of course the body needs some prostaglandin and that which is lost through taking an NSAID can be replaced by taking misoprostol. This is a synthetic analogue which prevents and promotes healing of stomach and duodenal ulcers caused by the NSAID. It is usually prescribed for the frail and elderly from whom the taking of an NSAID cannot be withdrawn.

If arthritis spreads to more joints of the body, then a disease-modifying anti-rheumatic drug (DMARD) will be prescribed and if no benefit results despite various ones being tried, it will eventually mean surgery to replace the worst affected joints. Joint replacement can transform the life of someone with osteoarthritis and it is now a common procedure, and can even be done for fingers. The hope for the future is that this stage would not be reached and that new medicaments will be found that will actually *cure* the condition and some drugs of this type are being actively researched. Until then we still have some remarkable molecules to bring relief. One of the early ones came as a result of the dedication of one man who believed in what he was doing despite orders from his superiors to stop – see box.

Only trained medical personnel can prescribe DMARDs like auranofin, but there are others such as chloroquine, methotrexate, penicillamine, and sulfasalazine. Chloroquine was originally developed as an antimalarial drug but was found to have some anti-rheumatic activity. Methotrexate acts on the immune system and was particular favoured by doctors for the treatment of arthritis in the 1990s because it could be prescribed on a long-term basis. Penicillamine is effective but needs to be taken for many months before its benefits appear. Sulfasalazine was specifically developed as an anti-rheumatic drug and

The Dogged Determination of Arthur Nobile

For more than 40 years prednisolone was successfully used for the treatment of rheumatoid arthritis. This had been patented in 1954 and it came as the result of research by Arthur Nobile at the drug company Schering. He started a project of his own and, when told to cease work on it by his supervisors, he simply ignored them. That was in the late 1940s and early 1950s. Then in 1954 a specialist in antibiotics, Fernando Carvajal, joined the company and realised that Nobile's work was ground-breaking and they started to cooperate. Soon Schering was producing prednisolone, and by 1974 it was recognised as one of the major advances in arthritis treatment of the previous 20 years. Prednisolone blocked prostaglandins and was excellent at suppressing inflammation in arthritic joints. It was also used in the treatment of eye infections and inflammatory bowel disease. Limits were eventually placed on its use because, like other systemic steroids, it has undesirable side effects and it interferes with the action of white blood cells.

first used in the 1940s. Many patients whose arthritis has been confirmed are given a combination of NSAID and DMARD from the onset of diagnosis. Even when drugs produce the desired result, to the extent that the condition appears cured, patients will still be advised to continue with the medication to avoid a relapse.

NSAIDs offer relief by blocking enzymes known as *cyclo-oxygenases* (COX) which produce the prostaglandins that trigger pain and cause inflammation – but not all COX enzymes need to be blocked to achieve this effect. In the early 1990s a group led by Daniel Simmons at Brigham Young University in Provo, Utah, found that there were two types of cyclooxygenases: COX-1 which is always present in the body; and COX-2 which remains dormant until activated by a signal from cells that they are under attack. COX-1 has several roles, one of which is to produce the mucus that protects the stomach wall from digestive acids. COX-2 triggers the local production of prostaglandins and these increase the sensitivity of surrounding pain receptors and dilate blood vessels to bring in more help. This knowledge came as a result of investigations spanning three decades and began with the work of John Vane (1927–2004). He was a key researcher in this area in the 1970s, and for his work he was rewarded with the 1982 Nobel Prize for Medicine, along with Sune Bergström and Bengt Samuelsson. He had proved that prostaglandins were the hormone-like chemicals which the body produces in response to damage at the site of attack and they signal that something is wrong. By controlling the formation of prostaglandins it is possible to control the pain and the inflammation, and make the condition easier to bear.

NSAIDs such as aspirin blocked production of both types of COX which is why this drug, useful as it is, can damage the stomach lining, leading to bleeding and ulcers. What was needed was a drug that blocked only the formation of COX-2. In 1992, research scientists at the Merck pharmaceutical company began to search for just such a drug and there were already some that appeared to work this way, including an organofluorine drug produced by Schering and called flosulide. The Merck chemists produced a particularly successful one, rofecoxib, which is better known by its brand name Vioxx. Tests were undertaken to see if Vioxx was safe to prescribe and it appeared to be so. In 1994 →drug trials were carried out on volunteers in Belgium and these showed that the drug was well tolerated and that its effect on COX-2 would last for 24 hours. In Texas it was tested on people who had had their molar teeth extracted and proved to be a good painkiller, kicking in after about 45 minutes. Yet more studies were undertaken and all indicated how good Vioxx was. Year-long trials involving 25,000 patients, average age 62, went ahead, and again the results were all that had been hoped for. A group of around 160 volunteers even agreed to take 10 times the highest recommended dose and then have their stomachs examined by an endoscope. This research showed that there was none of the stomach damage that aspirin can cause. In 2000 the results of trials on 8,000 patients were published in the prestigious *New England Journal of Medicine* indicating that Vioxx was better than other NSAIDs. Vioxx had been launched in May 1999 and it quickly became the world's fastest growing prescription arthritis medicine. Its success was not to last.

Double-blind tests were carried out on the long-term effects but the results were somewhat worrying. They showed that while Vioxx was well tolerated by the body, after about 18 months those taking it were slightly more at risk of having a heart attack or a stroke than those on a placebo. Blood contains a natural chemical thromboxane, which causes it to clot and this is made by a process which involves COX-1. Blocking COX-1 will therefore reduce the risk of a blood clot forming and causing a heart attack. For this reason aspirin is taken by millions of people as a way of thinning the blood and preventing this from happening. A drug which targets only COX-2 does not provide this extra benefit and consequently it increases the risk of having a heart attack or a stroke.

Despite these findings, the FDA arthritis advisory committee reported in early 2005 that the benefits of COX-2 inhibitors far outweighed the other risks, and they decided that Vioxx was safe but patients should be made aware that its use carried a risk of heart attack or stroke, and that it should not be prescribed to patients who are vulnerable to these conditions – and that included cigarette smokers and diabetics. Earlier, in 2001, the UK's National Institute for Clinical Excellence, which vets all drugs, had recommended that COX-2 inhibitors should not be given on a routine basis but be reserved for patients at high risk of developing stomach bleeding.

Merck had withdrawn Vioxx in September 2004 as a result of its long-term trial and there followed a welter of articles declaring Vioxx to be dangerous and that Merck was really to blame. Leaked Merck documents and e-mails suggested that the company knew of the long-term side-effect of Vioxx as far back as 2000. Such wild allegations were answered by Peter Kim, the President of Merck Research Laboratories, who pointed out that the safety of Vioxx had been vetted by the FDA in April 1999 and February 2001 and it had not been thought necessary to issue safety warnings at that time. Trials had been carried out on more than 30,000 individuals and it was only later, after a 3-year long study, that new data showed a risk of heart attacks, but that this risk only applied to patients who had been taking the drug for more than 18 months, and that it was only after 30 months that the risk became statistically significant.

As a result of these findings Merck took Vioxx off the market in September 2004, and the week it recalled all supplies of the drug saw the company's shares drop from $55 to $32 – a month later they were down to $27. What threatened the company most was the litigation coming from the partners of those who had died while taking Vioxx, and they were encouraged by some spectacular verdicts, such as one in Texas where a jury awarded an unbelievable $253 million in damages to the widow of a man who had died of a heart attack, while another jury awarded a mere $32 million for the death of 71-year-old Lionel Garza. Maybe that jury was somewhat influenced by the fact that he had been a life-long smoker and had already undergone quadruple-bypass surgery for heart disease some years previously, although they still blamed Vioxx for causing his death. In another trial the jury was unable to decide whether Vioxx had caused a fatal heart attack, which was an understandable verdict considering the man had only taken the drug for a month. More than ten thousand such law suits are now pending in the USA. However, the legal vultures did not have it all their own way. In November 2005 Merck won a court case in New Jersey against an Idaho man who had taken Vioxx and subsequently suffered a heart attack.

Vioxx may no longer be available, but the theory behind its development was scientifically sound and other drugs are coming to take its place, such as Prexige which was launched in January 2006 and which its makers, Novartis, say has been tested on 34,000 patients and shown to be no more likely to cause heart attacks or strokes than the other commonly used painkillers like ibuprofen.

Today we know a lot more about arthritis. The destructive stage of the disease is associated with cells in the immune system that release cytokines and growth factors into the blood stream and it is these which stimulate the tissue around the joints to begin their attack on the joint itself. What triggers the immune system is still a mystery. Some pharmaceutical companies have developed drugs that will block the formation of these cytokines and several of these are undergoing trials. One of the more important members of the cytokine family is TNF (tumour necrosis factor) which is secreted by white blood cells and has a key role in fighting infection, and it too joins in the attack on damaged cartilage. TNF is continually present in the blood of those with the disease, but two anti-TNF drugs are now available: infliximab (trade name Remicade) and etanercept (trade name Enbrel), which bind to TNF and remove it from circulation. A monoclonal antibody drug, Humira, produced by the biotechnology company Cambridge Antibody Technology and Abbott Laboratories, has also proved highly effective at blocking TNF and sales of this exceeded $1 billion in 2005.

As populations in developed countries continue to age then arthritis will become more of a problem. There is every likelihood that the cause of this disease will be discovered in the next decade and then the research chemists in pharmaceutical companies will begin to look for drugs that can tackle it at its very early stage and even stop it happening altogether. Meanwhile there is a natural chemical which is popular with those who want to ward off arthritis and that is glucosamine, taken by millions in the belief that prevention is better than a cure.

Glucosamine is a component of human cartilage and it is found in the cells of human joints. Many who suffer from arthritis, or wish to ward off this disease, can take it as a dietary supplement and it appears to help. Those who can afford to take it are expected to pay dearly for the benefits it brings. At least that seems to be the philosophy behind the exorbitant mark-up in price as glucosamine moves from food industry waste to local pharmacy shelves. More than 5,000 tonnes of glucosamine are manufactured and consumed worldwide every year. It is made mainly in China where it is extracted from the natural polymer →chitin taken from the shells of shrimps and other crustaceans. (Chitin is the second most abundant biopolymer after cellulose, and vast quantities used to be disposed of to landfill or dumped in the sea.) When chitin is treated with a strong acid, such as sulfuric acid or hydrochloric acid, it breaks the polymer down into individual glucosamine units. The bulk price of glucosamine is around €7 per kg, making the global production worth around €37 million. Of course when you buy it in health food shops the price is more likely to exceed €70 per kg or even more. Glucosamine may be a natural chemical but it too comes with unwanted side effects so it is not suitable for everyone – there is more about chitin in the next section. What may come as a surprise to many who take it is that glucosamine is basically a carbohydrate derivative,[18] which brings us to the subject of carbohydrates as agents of better living.

Carbohydrate as Cures

Most people regard carbohydrates simply as food components which provide only calories and from which our body gets its energy by digesting them with the aid of enzymes. They may also know that there are some carbohydrates which cannot be digested because they are impervious to these enzymes. Sugar is the best known example of the digestible sort of carbohydrate and cellulose of the non-digestible kind. Although cellulose consists of long chains of glucose molecules, it passes through the body as 'fibre' which nevertheless does some

18) Chemical formula $C_6H_{13}NO_5$. Glucosamine is glucose with an amine group (NH_2) attached to one of the carbon atoms. Glucosamine is often sold in conjunction with another amino sugar chondroitin which is involved in the fluid that lubricates the joints.

good by preventing constipation. Carbohydrates as food will be discussed in Chapter 4; here we look at their potential as medical treatments of the future. Research in this area promises to raise medicinal chemistry to a new level of understanding and provide cures for diseases we now think of as incurable. Nature creates and deploys carbohydrates in ways we are only just beginning to realise, let alone understand.

Carbohydrates appear to be involved in all kinds of ailments, including asthma and cancer. A group led by Jack Elias of Yale University, Connecticut, thinks that chitin might be the cause of asthma because he found that there is a lot of the enzyme *chitinase* in the lungs of asthma sufferers. This enzyme breaks down chitin, which is not a component of the human body but is an essential part of crustaceans, insects, and fungi, all of which can provoke an immune response in humans. *Chitinase* is a redundant enzyme in terms of human metabolism but maybe it invents a role for itself by stimulating the immune system to respond to the little bit of chitin that it encounters, thereby starting an asthma attack. If this theory is correct, then asthma might well be cured by disarming the *chitinase* enzymes and so Daan van Aalten at the University of Dundee, Scotland, began to screen known drug molecules for ones that could do this. To his surprise he found that caffeine blocked *chitinase*, which would explain why this natural chemical has long been used to control an asthma attack by relaxing the bronchial nerves. Van Aalten found an even better *chitinase* blocker is pentoxifylline, which is closely related chemically to caffeine. (Pentoxifylline was originally produced as a drug to increase brain blood flow by softening blood clots.)

Carbohydrates are the most abundant chemicals of biological origin, and they constitute around three quarters of the dry weight of plants. Carbohydrates have been somewhat neglected by medicinal chemists, but two things have recently happened that make them more interesting. First was the realisation that they are essential to the workings of cells and second was the discovery of new methods for making them in the laboratory. For centuries the only role for carbohydrates in pharmacy was the addition of sugar to disguise the bitter taste of some medicines. Then, in the 1930s, the complex carbohydrate heparin began to be used to prevent blood clots from forming and this is extracted from pig intestines. It consists of a chain of five carbohydrate units, and it can now be made synthetically thanks to work by the Australian company Alchemia, based in Brisbane, Australia. It deactivates blood clotting factors and is given after operations such as hip replacement and abdominal surgery.

It has been known for quite some time that carbohydrates have a critical role to play in plants and microbes, but they are now recognised as important to all living cells and that includes human cells. Carbohydrates come into play when sperm meets egg. When a sperm finds the egg it recognises a carbohydrate on the egg's coat and binds to it, and this act instantly triggers calcium ions to become active. They switch on an enzyme that forms hydrogen peroxide in the membrane and this cross-links the long chain molecules of the membrane, thereby making it impenetrable to other sperm. At least that is what biologist Gary Wessel of Brown University discovered happening in sea urchin eggs in 2004 and it is more than likely that the same happens in human eggs.

Carbohydrates attract water molecules, which is probably why they are to be found on the outside of cell membranes, but compatibility with water is only a minor part of what they do. Some carbohydrates are vital for passing subtle messages between cells. Sometimes the carbohydrates are attached to proteins which are also to be found on the surfaces of cells and these carbohydrates are there to protect the proteins from attack by enzymes. Essential though they are, carbohydrates can be exploited to attack cells and they are the route by which viruses and toxins gain access.

The natural chemical ricin, which occurs in the castor bean, is a deadly poison and it takes only one molecule to kill a cell. Its first action on encountering a cell is to bind to a carbohydrate on the cell surface and wait. It doesn't have to wait for long before the cell decides to investigate and takes the ricin molecule inside; it thereby seals its fate. The ricin migrates to the only site in the cell where proteins are made and once there it blocks it. Eventual cell death is ensured. The deadly toxin released by *E.coli* O157:H7 plays a similar trick to gain access to cells. Viruses also dock onto external carbohydrates and the influenza virus is particularly good at doing this. But what Nature does, humans can also do, and we can do it to defeat these natural enemies.

Carbohydrates may be the basis of future vaccines. Vaccines stimulate the immune system to produce antibodies against invading microbial pathogens. Vaccines are made from weakened or dead pathogens, or from proteins extracted from them. In theory, vaccines could be made to identify and react to the complex carbohydrates that protrude from the invading pathogens. While such carbohydrates have been chemically identified, this seemed an unrewarding route to vaccines because chemists were unable to make copies of the carbohydrate without going to extraordinary lengths, sometimes needing to carry out a sequence of more than 40 separate chemical reactions – and yields were tiny. In the past few years things have begun to change. Joining together two carbohydrates has always presented chemists with a problem, not because it is difficult to do, but because there are so many ways in which molecules can link up, giving a multitude of products of which only one is required. Identifying natural carbohydrates is not easy, but making them in the laboratory is much more difficult.

Complexity really is the name of the game when it comes to carbohydrate chemistry. Consider the simple glucose molecule. This consists of a ring of five carbon atoms and one oxygen atom with five reactive hydroxyl groups attached. It is through these groups that two glucose molecules can be joined together and there are 11 ways in which this can be done.[19] Add a third glucose unit to the molecule and there are now 176 possible combinations; add a fourth and there are 1056, and add a fifth and the number rises to an incredible 2,144,640. In the face of such complexity, the task of making a particular one is daunting, and yet the enzymes in a cell can assemble the exact carbohydrate it needs and often with ten or more carbohydrate units joined together.

The restrictions to making carbohydrates in the laboratory have been removed in the past ten years by some very clever research into methods of fabricating them. In the 1990s the group led by Steve Ley of Cambridge University, England, became pioneers in the synthesis of complex carbohydrates from simple carbohydrates. They found ways of joining as many as 11 of them together in a single reaction mixture. The skill in doing this was to find so-called protecting groups that can be attached to active sites of a sugar molecule to prevent these reacting while the desired ones then react as required. New methods of making simple carbohydrates have also been discovered, and these make it possible to construct a wide variety of new molecules, ones that even bounteous Nature herself does not make. David MacMillan

19) There is more about linking two glucose molecules together in Chapter 4.

and Alan Northrup, of the California Institute of Technology found that the amino acid proline works as a catalyst, allowing reactions to be accomplished at ordinary temperatures – as in Nature. They have managed to achieve in two steps a process that in the past has required as many as 44. Now it is even possible to make carbohydrates that do not occur in Nature and one day these may find uses in making new drugs.

The biggest hope for carbohydrate chemists is that anti-cancer vaccines will become possible. Cancer cells have alien carbohydrates protruding from their outer walls and when they move around the body they use these carbohydrates to attach themselves to other organs, thereby starting secondary cancers, a process known medically as metastasis. It might be possible to identify cancer cells by their carbohydrates, and if this could be done then a vaccine against such cancer cells might be possible. One such carbohydrate, found on prostate, colon, and breast cancer cells is globo-H and when this is injected into mice it generates antibodies which then recognise the tumour cells as foreign. Another way to attack cancer cells is to interfere with the enzymes they need to construct the alien carbohydrates and research on mice has shown that the size of a melanoma can be reduced by more than 99%, by inhibiting these enzymes. In 2003 Dextra Laboratories and Glycomed Sciences of the UK came up with a complex carbohydrate drug that has the remarkable ability to dissolve skin tumours.

The first carbohydrate-based vaccine to be made was approved in Cuba in November 2003 and is known as Quimi-Hib. It protects against *Haemophilius influenzae* type b (Hib) which causes pneumonia and meningitis in infants and young children. The vaccine came as a result of work by Vicente Verez-Bencomo and Violeta Fernández-Santana of the University of Havana. Hib-pneumonia vaccine now protects a quarter of a million Cuban children and whereas previously there were hundreds of cases a year, in 2005 there were only two. Cuba is a major sugar producer and naturally its scientists are motivated to seeking ways it can be used. Sugar is likely to become one of the main sustainable sources of the chemical industry in the future, and it is an abundant raw material. The sugar molecule is a combination of glucose and fructose and the bond between the two is easily cleaved by acid and they can then be separated. Glucose is the more important in terms of making pharmaceuticals. Glucose can also be obtained from cellulose but that reaction is chemically more difficult although it can be achieved by a combination of acids and catalysts.

An anti-malaria vaccine based on carbohydrate recognition might one day be possible. The first step towards this end was identifying the toxic carbohydrate that comes from the parasite and which causes the disease; the second step was to make the carbohydrate in the laboratory. When its synthesis was first attempted in 1995 it took five people more than two years to produce a few milligrams, but thanks to methods developed since then it can now be made in quantities a thousand times larger. Meanwhile other groups are developing carbohydrate vaccines to protect against fungal infections and staphylococcal-type infections such as cholera and typhoid. Indeed it may one day be possible to have various complex carbohydrates in one molecule so that the antibodies against several diseases will be formed from a single injection.

Carbohydrates also play a part in treating diseases by transporting drugs to where they are needed within the body. Often much larger doses of a drug have to be prescribed because it may not be very soluble in blood. An alternative approach is to wrap the active agent inside another molecule that is soluble in blood and so deliver the dose to where it is needed. Such a package molecule is cyclodextrin and this consists of six, or eight, glucose units that are joined together in a large ring rather like a wristband. While the outside of the cyclodextrin is water-seeking the inside cavity is much less so and indeed will accommodate a molecule that is water-insoluble or even one that is sensitive to reaction with water, and get it safely to its destination.

Carbohydrates are also proving to be beneficial to the human body in other ways, and especially the carbohydrate made by bacteria – see box.

Bacterial Carbohydrate

Bacteria make carbohydrates because, like plants, they need cellulose as a support material. The bacteria spin this polymer into thread-like strands and these weave themselves into ribbons. It has been said that bacterial cellulose is the Rolls-Royce of celluloses. What differentiates bacterial cellulose from plant cellulose is its strength and the way it retains water. This combination of properties attracted the attention of scientists at the University of Jena in Germany where research chemist Dieter Klemm and surgeon Friedrich Schiller have used bacterial cellulose to make artificial blood vessels for use in microsurgery. They call their material BASYC, short for *BA*acterial *SY*nthesized *C*ellulose.

Bacterial cellulose is also used to treat long term patients who develop bedsores, which are large ulcers caused by lack of circulation and which can persist for years. Xylos, a small company in Langhorne, Pennsylvania, now supplies a dressing for such wounds called XCell, made from bacterial cellulose. This not only reduces the size of the affected area but it promotes healing of the underlying damaged skin.

Another bacterial cellulose membrane, devised by researchers at the Technical University of Lodz in Poland, can be applied to severe burn wounds and it helps them to heal much quicker.

Anaesthetics[20)]

When there were no anaesthetics, surgeons had to work fast, and some could amputate a leg in less than 30 seconds. Some operations, such as removing stones from internal organs, took a little longer and all the unfortunate patient could do was writhe in agony while tightly strapped to the operating table, or held down by burly theatre assistants. There were crude pain-killers, such as alcohol and opium, and there are even reports of people being knocked unconscious with a blow to the jaw before an operation, but the idea of operating on a person deeply asleep only became possible with the introduction of chemicals whose specific action was to render them oblivious to pain as well.

There is a suggestion that in 1540 the great pharmacist Paracelsus (1493–1541) noted that ether rendered birds temporarily unconscious, but he experimented with it no further. More than 200 years were to pass before the next step was taken and this was the discovery of nitrous oxide gas by Joseph Priestley in 1772. Breathing this gas had amusing effects and it became popularly known as laughing gas. On 26th December 1799, the great scientist Humphry Davy tested it on himself by sitting in a closed box into which 20 quarts (about 20 litres) of nitrous oxide (N_2O) were introduced. He afterwards wrote:

20) The words anaesthesia and anaesthetic were first used by Oliver Wendell Holmes in a letter to the surgeon William Morton dated 21st November 1846. The Association of Anaesthetists of Great Britain and Ireland have a museum in central London, at 21 Portland Place, where 3000 objects used to relieve pain, and dating from around 1774 to the present day, can be viewed. Those intending to visit should first make an appointment (phone: 0207-631-1650) – admission is free.

"I lost all connection with external things. Trains of vivid images rapidly passed through my mind and connected with words in such a manner as to produce perceptions perfectly novel...Nothing exists but thoughts! The universe is composed of impressions, ideas, pleasures, and pains."

Little wonder then that it became a popular form of amusement, especially among students. Davy did more than enjoy the high of N_2O and he suggested that the gas might be made use of to induce a state of painless unconsciousness. No surgeon seems to have taken up the idea and it was to be many years before simple chemicals began to be tested on humans for this purpose. In the 1820s an English surgeon Henry Hill Hickman tried carbon dioxide as a way of making animals unconscious, which it did, but he never used it on his patients.

The first operation in which ether was used as an anaesthetic was carried out on 30 March 1842 by the American surgeon Crawford Williamson Long. In that operation he successfully removed a cyst. A few more operations followed but it is said that he was eventually accused of sorcery and threatened with lynching so, understandably, he went back to the old ways. Others were not so deterred and nitrous oxide was used on Horace Wells in 1844; he was a dentist who had one of his own teeth pulled while under its influence. Around the same time, at the Massachusetts General Hospital, surgeon William Morton successfully anaesthetised a patient with ether while he cut out a large tumour. In 1847 James Simpson showed that chloroform was also a good anaesthetic, and John Snow used it on Queen Victoria in 1853 while she was giving birth to Prince Leopold. She was duly grateful and with the royal seal of approval the future of anaesthesia was ensured.

The best anaesthetic vapour was found to be a 1:2:3 mixture of alcohol, chloroform, and ether, known as ACE. Despite being widely used, it was always known that ACE presented a risk, and yet it was the anaesthetic of choice for more than 100 years. Ether caused serious fires and explosions in operating theatres and while the risk of this was small – about one serious incident every 100,000 operations – it was alarming when it happened. Chloroform was not a fire risk but it could be deadly to some patients, killing them within minutes in certain tragic cases, and seriously damaging the liver of others. (Nitrous oxide was less risky, and continues to be used even today, but it does not produce deep anaesthesia.)

These early anaesthetics had been stumbled on by accident. Could chemists devise better ones? For almost a century the answer was no, although other chemicals were discovered that had anaesthetic capabilities and were experimented with, namely ethyl chloride, trichloroethylene, cyclopropane, and vinyl chloride, each of these had its drawbacks, not least because they too were highly flammable or had toxic side effects. What was lacking was knowledge of *why* chemicals like chloroform worked. Clearly an anaesthetic had to be volatile and easily absorbed from the lungs to the blood stream and thence carried to the brain. The molecules had to be the right size to block channels in membranes through which sodium atoms move, their motion being the way that electrical impulses move along a nerve fibre. Because membranes are fat-like chemicals then the more soluble an anaesthetic was in solvents that dissolved fats, the better it was likely to be. In addition, there were several other properties that the perfect anaesthetic should have: it must not be toxic; it must not be chemically reactive; it must not damage vital organs; it must not be flammable; it must be quick acting; it must have no unwanted after-effects such as vomiting; it should have a long shelf life; and it must not be too expensive. Both ether and chloroform have several of these desirable features and approaches to finding a new anaesthetic were either to make ether non-flammable or to make chloroform less toxic.

Chloroform had the advantages of volatility, non-flammability, and compatibility with fatty tissue, and while it is deadly to some individuals, most people are not damaged by it. Having fluorine atoms in a molecule rather than chlorines make it less toxic, as well as making it more volatile and this was the route to safer anaesthetics. The search for a better anaesthetic was undertaken in the laboratories of Imperial Chemical Industries at Widnes, Lancashire, by a group led by Charles Suckling. These researchers made many gaseous and volatile compounds that incorporated fluorine, chlorine, and bromine atoms on the basis that the fewer hydrogens in the molecule the less flammable it would be and the more likely it would have a volatility that would ensure a high enough concentration of the vapour to produce anaesthesia. They also began to understand how anaesthetics could be designed so as to deliver the necessary amount of the chemical to the brain without a lot of it penetrating other organs of the body where it might be metabolised to products that are damaging. Eventually they came up with halothane which was patented in 1958 and proved to be a success. Soon it was being widely used although there were signs that it was less than perfect, in that for some patients it affected their liver, where it was converted to the →carboxylic acid, trifluoracetic acid, which is toxic.

In the 1980s two more anaesthetics came into use: enflurane and isoflurane. These were not metabolised by the liver to the same extent, 2% in the case of enflurane, and only 0.2% of isoflurane. Enflurane was introduced into clinical use in 1981, but isoflurane was delayed because some research appeared to show that it caused liver cancer in mice. This research was repeated by others and shown to be wrong and isoflurane came into general use in 1984. What operating theatre personnel did not like was its off-putting smell. Were there health risks associated with the new anaesthetics? A statistical analysis of the side-effects experienced by 17,201 patients on whom they were used was compared with the effects experienced by a similar group on whom halothane had been used. Patients on the newer anaesthetics were more likely to suffer a heart attack and, with isoflurane, palpitations were more common. However, there was no increased risk of the patient dying.

Isoflurane has now given way to servoflurane, especially for operations on children. Servoflurane was first made in the 1970s, and it is a particularly powerful anaesthetic with rapid action and is the one preferred in Japan. A newer anaesthetic is desflurane. Both have low solubility in tissue so the anaesthetic does not diffuse into other parts of the body making recovery after an operation speedier and there is less memory loss. Desflurane is set to become the anaesthetic for day-case surgery.

Nitrous oxide is still used because of its rapid action and adding a little of this to other anaesthetics makes less likely another type of accident, that happens when the supply of anaesthetic runs out during the course of an operation and the anaesthetist is unaware of it. As a result, the patient starts to recover consciousness during the operation and consequently suffers extreme pain and trauma. Financial compensation in such cases can be particularly high. Nitrous oxide keeps the patient unaware of what is happening.

The safest anaesthetic of all is probably xenon, a gaseous element that exists as single atoms with very little inclination to react chemically. Its potential as an anaesthetic has been known for more than 50 years but it has never been widely used because of its rarity and consequent cost. Xenon constitutes only 2 parts per trillion of the Earth's atmosphere and while the total amount exceeds 2 billion tonnes, only one tonne of this is extracted each year and that is produced by liquid air plants. Why the Earth has so little xenon is an unsolved scientific mystery and it may be that a lot is locked away in the crust as gas hydrates or in rocks. Expensive though it is, xenon can have a highly beneficial effect in some operations, especially those involving coronary artery surgery, from which most patients emerge suffering some loss of their mental faculties due to the formation of microclots of blood. Collaborative research in 2003 between Imperial College London and Duke University Medical Center, North Carolina, showed that when xenon was used as the anaesthetic there was much less brain damage.[21)]

World sales of pharmaceuticals exceed $500 billion annually, of which the USA accounts for 45%, Japan 11%, Germany 6%, France 5%, and the UK 4%. China, the world's most populous country accounts for only 2%. The most prescribed category of drugs is the cholesterol-reducing ones which account for 6% of sales, followed by anti-ulcer drugs (5%), and anti-depressants (4%). The remaining 85% treat everything from stiff muscles to deep-seated cancers. Yet despite the health benefits that this industry has wrought and the human suffering it has relieved, there are many who distrust its products and suspect the industry's motives. Such people turn to alternative therapies and this brings us to one of the most popular ones: homeopathy.

21) Xenon may also have a future in medicine in magnetic resonance imaging (MRI) of the lungs. These cannot be seen by ordinary MRI but if they contain xenon then they can be seen and this is due to one of its isotopes, xenon-129. This accounts for 27% of xenon. It is activated by exposure to laser-stimulated rubidium atoms which transfer energy to the xenon atoms, thereby enabling them to be viewed by MRI.

Issue: Do Homeopathic Medicines Really Work?

The answer, somewhat surprisingly, is yes, and it is surprising because homeopathic medicines consist of nothing but water. They start as a solution of some active agent and this is diluted 99% with pure water. A drop of this solution is then diluted with a lot more water. A drop of that solution is further diluted, and so on for up to 12 times, or even 30, at the end of which it is possible to calculate that not one molecule of the active agent is still present. Strangely, the more you dilute the medicine, the more potent it is supposed to become, the potency being produced through a special way of pounding the solution known as succussion. (The father of homeopathy, Samuel Hahnemman, said that the vessel containing the solution should be struck 150 times against a leather-bound book. Today the succussion is done by machine.)

So how can a homeopathic remedy work? The answer is not in the medicine but in the lengthy diagnostic session with a sympathetic homeopath that precedes the treatment. It is this *psychotherapy* that benefits the worried patient. The homeopathic remedy then confers its benefits in the same way that a placebo often produces benefits in people when these are given in large scale trials of real drugs. Placebos are generally harmless and inactive substances such as cellulose, but they do produce positive results for around 30% of those who take them. And so it is with homeopathic medicines which do work for many people, but for unbelievers like myself they are a waste of time and money, because we realise that all we are drinking is water.

Some homeopaths claim that the water in their medicines has preserved a 'memory' of the active agent that was there originally. There have even been papers published in respectable scientific journals that purported to show that even after many dilutions the water still had the ability to affect living cells. Further research has invariably shown them to be wrong, although such doubts do not surface in those journals devoted entirely to homeopathy. The idea that water could preserve an imprint of a molecule that it once contained would answer the sceptics charge that there is nothing there, but then all water would contain memories of all kinds of things and not just those we think might be beneficial. Indeed it has been calculated that those who drink water in London are mainly drinking water taken from the River Thames which has passed through the pipes and sewers of many towns upstream. What molecules it might still retain in its 'memory' does not bear thinking about.

Assessing the effectiveness of alternative or complementary medical treatments is not easy. Those committed to scientifically proven treatments of the kind described in this chapter often dismiss them as little more than quackery even though they appear to work for some people. The subject is admirably dealt with in Toby Murcott's book *The Whole Story: alternative medicine on trial* published in 2005. He concludes that if the issue of whether they are effective is to be resolved then a degree of humility is going to be required by those on both sides of the debate.

3
Better Loving: The Not-so-dirty Weekend

A small arrow printed before a word in the main text indicates that there is more information on that topic in the Glossary.

News from the Future

Global Times News, 21 March 2025

Armpits Advance Thoughts of Romance!

A deodorant that works by converting the malodours produced by bacteria in the armpits into molecules that can make you feel romantic is being marketed by ChinaChem, the world's leading chemical company. Yesterday saw 'Love Chemistry' launched at the world's leading fashion centres: Hong Kong, Mumbai, Rio de Janeiro, and Tokyo – and at a midnight ball in New York.

Head of ChinaChem, Carlos Mao Tung, said that it represented three years of intense research by chemists at the company's Shanghai laboratories, followed by three years of safety and consumer testing. "Love Chemistry appeared to transform the lives of some people who tested it and we received glowing reports of its effects. Research revealed that it increased the level of serotonin in the brain, making people feel better about themselves. This creates a willingness to engage in sex and to enjoy it more".

Asked about the likely ban on Love Chemistry in the European Union, on account of the chemicals it contained, Mao Tung said he regretted that people there would not be able to benefit from the new perfume although it would be on sale in neighbouring non-EU countries such as Switzerland and Norway.

Page 5: Leading movie star tells how she tested Love Chemistry with the help of two male leads.

Page 6: Paris perfumer says Love Chemistry could make young women infertile.

We speak of the *chemistry* between lovers, using that word to signify strong mutual sexual attraction. Of course a lot of activity in preparation for such an encounter involves the other kind of chemistry and that is what this chapter is all about. The first part looks at ways we ensure that we are not putting potential partners off because of our body odour. The second part is about the way chemistry helps us to complete a successful sexual encounter, and the last topic is about the eter-

Better Looking, Better Living, Better Loving. John Emsley

ISBN 978-3-527-31863-6

nal problem of Nature's outcome for a sexual encounter not being the one we want. In other words we may only have intended to engage in sexual recreation but realise it might have resulted in sexual procreation. Then we need another kind of chemistry. Finally there is the issue of chemicals that are criminally misused to obtain sex: the so-called date-rape drugs.

Smelly Chemistry

The sense that seems to be particularly involved with sex is the sense of smell. It is both intimate and animal, and it can stir the deepest emotions, yet it is the sense we know least about, and in humans it is less well developed than in other species. Nevertheless, it is still very powerful and it is *the* chemical sense. Smell is the way we detect molecules, and our noses are capable of analysing them in a remarkable way. In the nasal cavity there is an area covering four square centimetres from which dangle millions of nerve fibres like tiny hairs. These have a film of mucus on their surface onto which molecules in the air are temporarily absorbed, either to be instantly recognised or, if it is an unfamiliar smell, to be identified as something attractive or repulsive. The receptors are so sensitively tuned that they can tell the difference between right-handed and left-handed molecules which only differ in that they are mirror reflections of each other, a property of molecules known as →chirality. Thus they can differentiate left-handed limonene, which smells of pine, from right-handed limonene, which smells of oranges.

Just how sensitive we are to smells can be seen by comparing our perception of certain substances. Ammonia gas is regarded as very smelly and we can detect this if the level in the air we breathe rises to 5 parts per million (ppm) which is the same as 5000 parts per billion (ppb). This latter unit is more useful because we can detect some smells in tiny amounts, such as the bad egg smell of hydrogen sulfide at 100 ppb, the bad breath molecule dimethyl sulfide at 1 ppb, and the excrement smell of skatole at 0.4 ppb.[22] While ppb is not a familiar

22) Although this molecule gives human excrement its characteristic odour, it is also a chemical used in flavours and fragrances in tiny amounts. At low concentrations it imparts a pleasant, sweetish aroma and it also is a natural component of the fragrance of the arum lily. It chemical formula is C_9H_9N.

way of measuring things, it is possible to think of it in terms of time: 1 ppb is like 1 second in 30 years, which means that 0.4 ppb is like 1 second in a lifetime of 75 years.

The molecules whose smell we generally find most offensive are →amines, →carboxylic acids, and →sulfur compounds, but it partly depends on the situation in which we encounter them. Take the carboxylic acids, for example. Acetic acid (vinegar) is an unpleasant aroma when associated with a wine because it tells us the wine is going off, whereas we may find it attractive when it is sprinkled on fish and chips. As the chain of carbon atoms attached to a carboxylic acid gets longer the smell begins to change. Thus when there are four carbon atoms we have butanoic acid which is the smell of rancid butter, and when there are six atoms in the chain we have caproic acid which takes its name from *caper,* the Latin for goat. Goat's cheese contains this acid and then we may enjoy it, but it is also the odour of sweaty feet when we deem it very unpleasant.

The human body can give off odours that somehow seem designed to repel. Thus while fresh sweat is inoffensive, and may even be somewhat attractive, it soon becomes rancid due to the action of bacteria which generate acids including 3-methylbutanoic acid, which is also present in most cheeses made from cow's milk. Urine too is not offensive when it first leaves the body – indeed some of the deepest notes in perfumes even echo its odour – but eventually it will start to smell as bacteria get to work on it and then it emits amines. Bad breath likewise comes from the action of bacteria in the mouth working on the protein in food residues. In this case the offensive molecules are sulfur compounds such as methyl mercaptan and dimethyl sulfide. Methyl mercaptan can be detected by the human nose at concentrations in air of only 0.2 ppb. You also encounter these molecules if you eat the delicacy known as truffle, but then there is only a faint hint of them. These rare fungi live underground but pigs and dogs are capable of detecting their smell and are used to locate them.

Adding more sulfur atoms to simple molecules like these can intensify the smell even more. The infamous lily titan arum (*Amorphophallus titanum)*[23] blooms only rarely but when it does it may even make national news, as it did in California in June 2003, when it attracted more than a thousand visitors, and in the UK in April 2005

23) Also known as the corpse flower, it grows up to three metres tall.

when the plant at Kew Gardens blossomed forth. In May 2006 it bloomed in the botanical garden of Bonn University, Germany, and in a unique fashion, in that it produced several flowers simultaneously. Its smell has been described as that of rotten eggs laid by a decaying chicken in a blocked sewer. When this lily blooms it can be smelt a kilometre away but thankfully it flowers for only two days. This lily emits dimethyl disulfide with two sulfurs at the centre of the molecule, and dimethyl trisulfide with three sulfurs.[24] These molecules are the ones that attract flies to a dead carcass and the titan arum lures them to itself and holds them for a day before releasing them to go off and find other flowers to pollinate, again attracted by the sulfide molecules. Female golden hamsters also attract a mate by releasing dimethyl disulfide, but while that may arouse his curiosity it is not enough to arouse him sexually. She then has to release another agent, this time a small protein molecule, before he will mount her.

Insects and animals rely on smell to attract mates. The male emperor moth can sense the presence of a female even though she may be five kilometres away, which gives some indication of how few key molecules his sensors need to detect her. The scientific name for sex attractants is pheromones and they can be made in the laboratory and then used to lure some pests into traps, although these may catch only the males, and it only needs one male to find the female for successful breeding to occur. (I used such a trap to try and reduce the effect of the codling moth on the apples growing in my garden and while it caught dozens of males, it did not put an end to the problem.)

It is not always the female that gives off pheromones. The musk male deer attracts his mate by means of the chemical muscone which he releases from a special gland, and male pigs emit androstenone which turns female pigs on. Humans also produce this molecule in their armpits, males especially so, and it has a very faint musky odour. Could it attract women? Some men believe it could and sex shops sell this chemical in spray form. There is as yet no scientific evidence that it is a human pheromone, but there are suggestions that we may detect and respond to molecules even though we cannot smell them.

The pheromone detector in the human nose became redundant when humans opted to use other ways of attracting a mate such as using visual cues, and indeed our primates predecessors appear to have

24) Their chemical formulae are CH_3SSCH_3 and CH_3SSSCH_3 respectively.

lost interest in pheromones when they started to see in colour, which was 23 million years ago. Walking upright forced this evolutionary change but while it improved our vision it lifted our nose from the area where it would be most useful, which is nearer the ground. Others think that we may still have residual pheromone ability because research has shown that odourless molecules given off by men can indeed affect women. In 2003 a paper published by George Preti and Charles Wysocki of the Monell Chemical Senses Center, Philadelphia, reported that odourless secretions wiped from clean male armpits had two effects on women: in some it induced a feeling of calm, while in others it appeared to affect the hormones that govern ovulation.[25] The molecule responsible is thought to be androsta-4,16-dien-3-one, which is the main androstenone on male armpit hair and is also present in semen. Scanning of the female brain shows that the smell of this molecule can change glucose levels although what this means is not clear.

In 1986 the largest ever smell survey was carried out involving 1.5 million people from all over the world. They were given 'scratch and sniff' cards and a questionnaire to answer. The scents tested were androstenone, isoamyl acetate (banana), glaxolide (a synthetic musk), eugenol (cloves), rose, and a mercaptan (the one added to natural gas to make it detectable). Of those who took part, 50% were able to detect all six smells and only 1% could detect none of them. Women were found to be better than men. The smell that was sensed the least was androstenone, and if this really is a human pheromone then it does not operate by having an irresistible odour.

Three hundred years ago anatomists observed that humans had two tiny dents on either side of the nose cavity and about a centimetre up from the nostrils. This is the vomero-nasal organ (VNO), also known as Jacobson's organ, named after the man who published a detailed description of animal VNOs in 1811. The human VNO has generally been ignored because it was clearly of no importance since there are no nerve connections between it and the brain. It appears to be merely the vestige of some early evolutionary organ. The VNO is very impor-

25) Women may also give off molecules that affect those around them and especially other women. This would appear to be the explanation of why women living together tend to menstruate together even though they menstruated at different times before joining the group. What molecule causes this to happen has yet to be identified but research at the University of Chicago in the 1990s showed that it was linked to molecules given off by their armpits.

tant to other species and in mice it has a profound effect on their behaviour because if some of its protein structure is removed then they behave very differently, appearing to be less aggressive when competing for mates and less protective when nurturing their offspring.

In 1997 David Berliner, an anatomy professor turned entrepreneur, reported that when tiny amounts of a synthetic steroid were put on the VNO of men it made them relaxed and their heart rate and breathing rate slowed down. Clearly the VNO is not as inactive as we imagine. And searches for a human genome by a team at The Rockefeller University in New York in 2000 revealed that we still have five potentially functional genes that we share in common with the VNO pheromone receptors of mice. A human sex attractant may yet be developed and maybe sooner than we imagine because a great deal is being discovered about sexual chemistry – see box.

A Chemical to Turn Us On?

In June 2004 James Pfaus of Concordia University, Montreal, published a paper in the on-line edition of the *Proceedings of the National Academy of Science USA* announcing that the a protein molecule coded PT-141 not only causes erections in men but also sexually excites female rats. When it was given to these rodents they became frisky and used various dodges to try and induce male rats to chase after them and mate with them. Whether it would have a similar effect on women has yet to be identified. PT-141 is a hormone-type molecule that targets a receptor in the brain which is known as melanocortin and is linked in humans to sexual desire.

For now we are stuck with more conventional ways of manipulating body odour and the main concern is that we should not be emitting off-putting smells. We cannot tell how we smell to other people so what we tend to do is ensure that our skin is clean and that areas that might emit stale odours are dealt with. The next topics in this section are about the chemicals we use to clean ourselves, and then we will consider antiperspirants and deodorants.

Wash All My Cares Away

The best way to ensure we are not radiating unpleasant body odours is to wash ourselves. The easiest way to do that is to have a shower, and there are hundred of products that promise not only to clean us but to make our skin feel better. Our skin is protected by an oily layer known as sebum. When we take a bath or a shower we want to wash this layer away together with all its dead skin cells, dirt, and the bacteria that live on it.

Soap molecules have the requisite chemical components for solubilising greasy dirt, but as we shall learn in Chapter 6, a drawback with soap is that it forms an insoluble soap scum in hard water. On the other hand, synthetic surfactants do not form a scum, but these can be rather harsh in that they also wash away all the skin's protective oils leaving it feeling dry and itchy. It was for this reason that the first commercially successful surfactant, sodium alkyl benzene sulfonate, could not be used for personal products. But improvements were on their way. If the benzene component was replaced with a 12-carbon chain then it was much gentler and ideal for washing skin and hair. The name for this 12-carbon chain is →lauryl. However, sodium lauryl sulfate has a drawback in that it is not very soluble in cold water, producing a cloudy solution which does not look attractive when sold in a transparent bottle, although it is used in products like toothpaste and shaving cream. The answer is to put an →ethoxy group in the molecule next to the sulfate group. The product, known as sodium laureth sulfate, does not go cloudy in cold water and it is now the main ingredient in shampoos.

Despite the proven safety of these products as gentle surfactants, there have been campaigns against them – see box.

Dirty Tricks Played on a Clean Product

In the late 1990s various web sites, seemingly set up by voluntary consumer groups, started claiming that sodium lauryl sulfate and sodium laureth sulfate were dangerous. They said that these chemicals caused hair to fall out, babies to go blind, and cancers to appear. Some said these effects were caused by traces of nitrosamines, others said that dioxins were to blame. Neither claim was true, and indeed the belief that shampoos were contaminated by dioxins came as a result of a lack of basic chemical knowledge. Dioxane is a solvent used in the manufacture of these surfactants but it looks as if it were a misspelling of the word 'dioxin' which does refer to a class of definitely harmful toxins. (Dioxane has a pleasant odour but its vapour is dangerous.)

Visitors who were worried by the alarms on these web sites were directed to companies selling 'natural' products, which by implication were safer.

In Chapter 6 we will look more closely at surfactants, of which there are four kinds. All have two features: a hydrocarbon chain attached to a water-soluble end group. This end group is what differentiates the four kinds of surfactant according to whether it is positively charged, negatively charged, both positively and negatively charged, or not charged at all. It is the last two of these categories that are used in personal toiletries. Those with both positive and negative groups give lots of lather and do not sting if they get into the eyes and are used in baby shampoos. The surfactants with no charge are regarded as the gentlest surfactants of all and are used in cosmetics.

If it is to be used as a bubble bath, a surfactant has to be crystal clear and remain so when added to the water of the bath. Non-ionic surfactants are ideal, and they can hold the fragrant oil in suspension until this can be released as the emulsion breaks down and the oil eventually ends up attracted to the bather's skin. In a shower gel the main active ingredient may be sodium laureth sulfate or it may be one of the newer surfactants known as alkyl glycosides which are made from renewable resources such as sugar and plant oils. These are not only seen as gentle to the skin but they foam well with a creamy feel to the lather, which is what users like. These surfactants also enable manufacturers of such products to produce clear solutions sold in transparent bottles, which is what purchasers like to see in the belief that the contents are pure and gentle. Other ingredients in such products are sodium chloride (salt) which clusters the surfactant molecules together and makes the liquid much more viscous, a mild acid such as the →carboxylic acid lactic acid which lowers the pH, together with its sodium salt, sodium lactate, which then keeps the pH constant. An antibacterial agent also needs to be included to keep the product free from microbial growths and the ones generally used are the →parabens although these have led to unwarranted concerns among environmentalists. Finally there will be various fragrance oils and maybe even some tea tree oil which can produce a tingling effect on skin.

You've had your shower, you feel clean, and now you are ready to play the game of life. But what if the weather or venue is hot and humid? Then you might undo all your good work by perspiring heavily and causing wet areas on your skin and clothes, and in certain areas like the armpits bacteria will thrive. As they do so, they might well begin to emit the unpleasant odours we associate with sweat. The result is hardly the 'come closer' signal we want to be giving off. Of course you could disguise your natural body odour with an overpowering perfume, itself a bit suspicious. The best way of preventing your sweat glands from letting you down is to use an antiperspirant or a deodorant. What chemicals do these involve?

It's the Pits, Man

They say that eight out of ten men in the West now use some form of underarm product every day, the object being to protect against wetness and odour. Women have been users of these products for much longer, the first appearing more than a century ago, but in those days they were fiddly to apply, left a sticky residue on the skin, and could stain clothes. Today they are easy to apply, feel dry on the skin, and they don't leave white marks, even on black garments. Aerosol sprays have the advantage of quick application – a two second spray is all that is needed and the product is dispersed widely and evenly – whereas roll-ons take time to dry. Sticks are the most economical and while they are drier than roll-ons they can be uncomfortable to apply on dry hairy skin.

Antiperspirants rely on aluminium and zirconium compounds for their action and these can leave white marks on clothing, but this can be overcome by ensuring the particles are so small they are virtually invisible. These ingredients are perfectly safe and conform to regulations issued by the FDA in the US, and comply with the Cosmetics Legislations of 1976 and 1995 in the EU. These guarantees of safety have not prevented some people from claiming that they are a health risk, as we shall see. Nor were they the ingredients of most concern. It was the propellant gases used in aerosols that were far more dangerous – not to humans directly but indirectly, by their damage to the upper atmosphere. These gases were the CFCs (short for chlorofluorocarbons) and they were present in almost all antiperspirants and deodorants in the 1960–70s. They fell out of favour in the 1980s when it was realised how damaging they were to the Earth's ozone layer, and it is this which protects life on the planet from the harmful ultraviolet rays of the sun. Although alternative propellants are now used, aerosols have not regained their dominant position, nor are they ever likely to because even the current propellant gases, which are propane and butane, can also damage the environment by acting as greenhouse gases, as well as being dangerously flammable. People now prefer to apply antiperspirants and deodorants as either solid sticks or liquid roll-ons.

Skin has lots of sweat glands over most of the body but especially on the forehead, palms of the hands, soles of the feet, in the armpits, and around the groin. They are triggered to release sweat by heat or strong emotions and, inexplicably, we sweat more under our left arm than under our right arm. We each have something like 3 million sweat glands and they can, if necessary, excrete as much as five litres of water per day. There are two kinds of gland: the eccrine and the apocrine. The former emits a 1% solution of sodium and potassium salts, along with a few other chemicals, but this kind of sweat does not smell, and its role is to cool the body down by evaporation. These glands do not develop until puberty. The apocrine glands are found in the armpits,[26] groin, and the soles of the feet, and they emit a more complex mix of sweat which contains proteins and oils, including steroids and cholesterol. This kind of sweat is an ideal medium on which bacteria can feed and it is these microbes which break down the natural chemicals into the obnoxious odours that comprise the smell of stale sweat, among which those of the →carboxylic acids are most recognisable.

The chief culprits in generating armpit odours are the bacteria *Corynebacterium xerosis* and *Micrococcus luteus*, with *Staphylococcus epidermis* and *Staphylococcus aureus* playing minor roles. There can be as many as ten million bacteria cells per square centimetre of armpit skin compared to only 1,000 on the skin of the forearm, and this is as true for women's armpits as for men's, and yet the odour women give off is different because it lacks some of the ingredients that male sweat contains. Male underarm odour has three components: an acrid one, a musky one, and a pungent one. The first of these comes from short-to-medium chain acids, the second from steroid type molecules, and especially androstenone, and the third from sulfur-containing molecules.

The Roman poet Catullus (84–54 BC) poked fun at men whose armpits smelt like goats, and the compound they are giving off is 4-ethyloctanoic acid. This chemical evokes a strong response in female

26) Each armpit has around 25,000 sweat glands.

goats when they are on heat and it is used by male goats as an attractant. The effect on human females is just the opposite, it repels, and women can detect it at concentrations as low as 2 ppb in the air. The acid which most characterises body odour is 3-methyl-2-hexenoic acid, but male armpits can also emit another pungent acid, isovaleric acid. The steroids in sweat are androstenone and androstenol – as a pure compound, the latter is said to have the rankness of stale urine. Sulfur-containing molecules are only minor components of armpit odour but these have been detected, and in 2004 a team led by Anthony Clark at the Swiss fragrance company Firmenich was able to identify eight sulfur-containing alcohols of which one, 3-methyl-3-sulfanylhexan-1-ol, was particularly repulsive and smelt like onions. The culprits making it were the bacteria *Corynebacterium* and *Staphylococcus*.

The bacteria which generate sweat odours can be tackled in three ways: we can deny them the soup they feed on, or we can kill them off, or we can prevent them from converting sweat into smelly molecules. This last approach means finding ways to block the enzymes they use to make the smells and while certain metals, such as zinc, can do this quite well and are incorporated into some products, this is not really as effective as either starving the bacteria or killing them off. Antiperspirants are based on the first premise, deodorants on the second. Each method has its advantages and disadvantages as we shall see. Even so, there are some unfortunate individuals who cannot be helped by either type of product – see box.

Fish Odour Syndrome

In Shakespeare's play *The Tempest,* the slave Caliban is described by Trinculo the jester as follows:

> *"What have we here? A man or a fish? Dead or alive?*
> *A fish. He smells like a fish; a very ancient and fish-like smell..."*

The Indian epic poem *Mahabharata* of 400 AD tells the tale of the beautiful Satyavati who suffered from fish odour syndrome and was a social outcast, condemned to work as a ferry woman. However, one day a holy man fell in love with her and by a miracle changed her body odour to an alluring perfume. They lived happily ever after ... or so the legend goes. In fact Caliban and Satyavati were suffering from a genetic defect that even today condemns its victims to smell of rotting fish if they eat too much protein and then start sweating.

One of the smelliest of chemicals which the human body has to dispose of is trimethylamine, and we can detect the smell of this volatile molecule at levels of only 1 ppb. It is formed from the choline that is a component of cell membranes. Foods rich in choline, such as fish, give off trimethylamine when they start to rot and indeed when we digest fish our body also has to get rid of trimethylamine. It does this by transporting it to the liver where an enzyme, *monoamine oxidase*, adds an oxygen atom to it, thereby removing its smell, and it is then excreted in the urine. A little may end up being excreted in sweat.

Sadly some people lack *monoamine oxidase* enzymes and they may lead a miserable life because they easily emit a fishy odour. Any physical exertion or hot weather which makes them sweat soon has those around them quickly moving away. Indeed many victims feel isolated because relationships inevitably break down, and so they try to avoid human contact and often work at home. One person in 10,000 has the faulty gene that is to blame and this was identified in 1999 by Ian Smith and coworkers at Queen Mary College in London. His advice to sufferers is to avoid all fish, red meat, eggs and soya, and eat mainly chicken and salads as the foods least likely to generate trimethylamine in the body.

Antiperspirants

Naturally we don't want to be releasing malodours, or even appear to be sweating, when we wish to be in close proximity to other people and so we use an antiperspirant to suppress the apocrine glands. Sometimes the products that do this even have benefits that were not expected. In one study, antiperspirants were tested on the feet of new recruits at a US military academy who were then sent on daily 20 km marches. The result was far fewer foot blisters – as well as less smelly feet.

The active agents used in antiperspirants are aluminium and zirconium salts. These are designed to block the sweat glands, which they do by reacting with water to form a sticky gel of aluminium hydroxide or zirconium hydroxide which plugs the pores, and we know this is how they work because these tiny plugs have been observed by means of transmission electron microscopy. By definition, an antiperspirant in the US must reduce perspiration by at least 20% in half of all users, but this requirement is easily met.

The first commercial antiperspirant was Everdry, launched in 1902, and it was dabbed under the armpits using cotton wool. It was cold to apply, took time to dry, and was acidic, so that it irritated the skin and rotted clothes – but it was better than nothing. It contained →aluminium chloride hexahydrate and later →urea was added to reduce the acidity. In 1934 Arid Cream was launched and it had been developed by John H. Wallace, a chemist from Princeton, who worked for Carter Medicine. It relied on aluminium sulfate which was less acidic. Then in 1947 two chemists T. Gorett and M.G. deNavarre experimented with so-called *basic* aluminium chloride, which is not acidic, and is made so by replacing most of the chloride with hydroxide (OH). The new chlorohydrate antiperspirants reduced perspiration by 40%, and they are still common ingredients. In 1978 Dr Nathan Brown working for Unilever at its Port Sunlight research centre near Liverpool, England, found a way to make much smaller and more active aluminium chlorohydrate particles and these became the basis of Sure antiperspirant that was launched the following year.

In 1950 a chemist, H.L. Van Mater, patented antiperspirant formulations with zirconium chlorohydrates instead of aluminium chlorohydrates. They were even better than their aluminium counterparts but were more expensive. At the time it seemed that zirconium salts would be little used, but that changed when it appeared that aluminium was linked to Alzheimer's disease (AD). In the 1970s aluminium had been shown to be the cause of dialysis dementia, which afflicted patients on the early types of kidney dialysis equipment and who began to talk illogically, became forgetful, and seemed totally confused. The aluminium was being dissolved from the dialysis equipment and depositing in their brains. Another more serious finding claimed high levels of aluminium were present in the brains of those who had died of AD, and who were not on dialysis. The implication was that this metal was causing AD and an anti-aluminium campaign began. Eliminating it was not going to be easy because it was used in so many ways from cooking foil to indigestion tablets. (I threw away a perfectly good aluminium pressure cooker, thinking I was protecting my family, and bought a stainless steel one.) Even domestic water supplies contained traces of aluminium because aluminium sulfate was used on a large scale to remove impurities. While this use could not be ended easily, other more personal uses could be targeted and that included antiperspirants, not that the aluminium in those products penetrated into the body, although some anti-aluminium campaigners seemed to think that it did.

One answer was to reformulate antiperspirant products and replace aluminium with zirconium, but the need to do this was undermined when, in the 1990s, the re-analysis of brain samples taken from AD victims showed aluminium was not present after all; the 20-year scare had been based on faulty analysis. The new analysis carried out at Oxford in 1992 was confirmed in Singapore in 1999, and published in the highly-regarded science journal *Nature*. Aluminium is no longer thought to be the cause of AD. The anti-aluminium lobby were confounded and their campaign ended in disarray after a group of volunteers in 2000 agreed to take large doses of aluminium hydroxide for 40 days, during which time their urine was monitored for aluminium. This showed that they were excreting the metal at more than ten times the normal rate, and in some cases at 20 times the rate, but that this was having no effect at all on the body's immune system. In fact aluminium is well tolerated by the human body, which is not surprising considering it is an abundant part of the natural environment, being one of the main constituents of clays and soil.

Zirconium appeared to have lost out in the antiperspirant battle but it was not going to go away because it is better than aluminium. Today a mixture of aluminium and zirconium compounds is present in many antiperspirants, with some formulations having 10% zirconium and 90% aluminium whereas in others there may be as much as 30% zirconium. The aluminium and zirconium salts are produced as powders although they can be made in the form of large spherical particles containing propylene glycol and this enables them both to dissolve and appear transparent when in a dispersant medium. Clear roll-on antiperspirant is made by dissolving aluminium zirconium chlorohydrate in a mixture of water, alcohol, and propylene glycol, with silicones added to make the liquid viscous. A solid antiperspirant will have around 25% of the active agent in a mixture of silicone and a wax such as stearyl alcohol.

Attacks on antiperspirants have been fuelled by reports on the internet saying that they cause breast cancer. In 2002 a paper in the *Journal of the National Cancer Institute* refuted these accusations and reported that there was no indication that antiperspirants (or deodorants) had been more heavily used by women with breast cancer. This study by Dana Mirick and colleagues at the Fred Hutchinson Cancer Research Center of Seattle compared 813 women who had been treated for breast cancer with 793 women who did not have the disease, with care being taken to ensure that the two groups matched in terms of age profiles and other factors. In fact there is no scientific evidence to show how aluminium from deodorants could possibly gain access to the body, and indeed it gets no further than the plugs that block the sweat glands on the surface of the skin.

Other activists attacked deodorants from a different angle, basing their claims on the parabens which are added to them as preservatives, and linking these to breast cancer because tissue analysis showed minute traces of them to be present. The finding of parabens in breast tumours does not mean they *cause* the tumours – they don't – but in any case the value of the research was called into question because those who did the analysis forgot to check whether traces of these compounds also occurred in other, healthy, tissue of the body. Nor can material applied under the arms reach breast tissue, because it is not physiologically possible. The US National Cancer Institute said it was not convinced either that deodorants were in any way connected to this disease. (The FDA also said it had no evidence that these products caused cancer of any kind.) In another survey, 437 breast cancer sufferers were asked when they started shaving under their arms and using these products and it concluded that the younger you started, the younger you were when you got the disease. Again this research is unlikely to be verified.

Deodorants

Thorough washing of the armpits can remove 99% of the bacteria that are there but the remaining 1% can multiply rapidly and within a few hours there are millions of them again, digesting sweat and releasing offensively smelling molecules. An obvious way to prevent this is to kill as many of them as possible, and a typical deodorant will contain around half a per cent of an antibacterial agent such as →triclosan. The first deodorant, Mum, came on the market in the US in 1888 and consisted of lanolin cream with zinc oxide as the active agent. Mum dominated the market until the 1960s when Gillette introduced an aerosol spray deodorant, Right Guard, which contained not only a zinc compound (zinc phenolsulfonate) but hexachlorophene,[27] which is a powerful antibacterial agent. This chemical was banned in the mid-1970s because it is toxic, and the ban came as a result of a French company putting 6% of hexachlorophene into a baby powder in 1972,

27) Its chemical name is 2,2-methylenebis(3,4,6-trichlorophenol) and its chemical formula is $C_{13}H_6Cl_6O_2$.

which was eventually responsible for the deaths of 30 babies. What was needed was a perfectly safe antibacterial agent and that turned out to be triclosan. This can be incorporated into nanospheres, which means it is released over a longer period of time and so exerts better control over bacteria for longer, as well as reducing the likelihood of any adverse reaction on the skin. Triclosan is used in toothpastes, hand wash liquids, and deodorants, and it does the job for which it was designed – killing bacteria – supremely well.

Triclosan has not been immune to attacks which are to be found mainly on web sites. Most such sources are less than scientific and far from objective, so can generally be ignored, but a more serious attack was that by the World Wildlife Fund (WWF) in July 2005 when they reported that those who washed dishes using a detergent that included this antibacterial agent were exposing themselves to 'significant' levels of chloroform which it formed in water; 'significant' in this case being measured in parts per billion. The WWF said that chloroform had been 'linked' to bladder cancers and miscarriages, and they deduced this threat from a paper written by Krista Rule, Virginia Ebbett and Peter Vikesland, of Virginia Polytechnic, and published in *Environmental Science and Technology* (2005), but they jumped to the wrong conclusion. What that paper actually said was that chloroform was produced not from the triclosan or the hypochlorite ($HOCl^-$) that is used to treat water supplies but from *free* chlorine (Cl_2) the amounts of which are barely detectable.

A paper by Christopher McNeill and William Arnold of the University of Minnesota found that triclosan could produce dioxins under the influence of sunlight and they reported this fact in *Environmental Toxicological Chemistry* (2005). Again the amounts of dioxins were miniscule and so low that they could not possible present a risk to anyone who used an antibacterial washing-up liquid or who went swimming in a pool that contained traces of triclosan from the bodies of those who had used a deodorant and then swam in that pool. Triclosan is targeted by environmental groups merely because it is an organochlorine compound – in other words it has a carbon-to-chlorine chemical bond – and as such was thought to be unnatural. Of course some organochlorine compounds do cause cancer but being an organochlorine compound does not automatically make something carcinogenic. Indeed we now know that there are more than 4000 *natural* organochlorine chemicals including ones produced naturally by our own bodies.

Proper research into finding better deodorants continues and new findings come to light. For example, underarm odour can also be reduced by applying derivatives of lactic acid and especially those with 12 or 13-carbon chains attached. Why this should work is somewhat puzzling but it has been suggested that the microbes that cause underarm odour prefer to consume this material in preference to natural body oils and in doing so they form non-odorous by-products instead of the usual malodorous compounds.

Something for the Weekend, Sir?

Condoms were once very much frowned upon in the UK and only obtainable from pharmacies, and then only by asking the pharmacists for them because they were not on display. This somewhat embarrassing encounter was solved by hairdressers who would ask their clients the seemingly innocent question: "... and something for the weekend, sir?" This coded message simply meant "do you want some condoms?" and if the answer was positive the barber would surreptitiously sell you some. Today condoms are on sale everywhere and they are deemed an essential item in men's wallets and women's handbags, there ready for use if their owners want to engage in recreational sex. The primary object now is not to avoid an unwanted pregnancy but to prevent sexually transmitted diseases.

Condoms have a long history[28] and there is even a suggestion that an ancient Egyptian inscription shows a man wearing one, but as he is not in the erect state it may represent something else. The same can be said of a picture drawn on the wall of a cave at Combarelles, France, and dated to around the time of the Roman Empire. The first true condom was devised by Gabrielle Fallopius (1523–1572), Professor of Anatomy at Padua University. (He specialised in studying the female reproductive system and he coined the word vagina.) Although he wrote nothing about his condom during his lifetime, his notes were published after his death, and they reveal that it was more like a cap

28) The largest collection of condoms in the world is held by Amatore Bolzoni of Italy who, according to Guinness World Records 2003, has 1,947 different kinds of condom.

rather than a sheath and it was held in place by the foreskin. He had designed it as protection against the new disease syphilis which was then spreading throughout Europe and which, in its early years, was particularly virulent. The Fallopius condom was made of tightly woven linen and it may have been treated with various liquids with the object of making it better at preventing infection. His notes show that more than 1000 men were instructed on its use but clearly it did not catch on.

There was no written record of condoms for another hundred years until one is described in the anonymous book *L'École de Filles* [School for Girls] which circulated in Paris in 1655. This described a condom made of linen, the object of which was to prevent conception. The French upper classes at the time of King Louis XIV generally preferred to use a female contraceptive in the form of a sponge inserted into the vagina.[29] Meanwhile the British had been making condoms from animal intestines, five of which were found preserved in a latrine at Dudley Castle in Staffordshire. This had been a royalist stronghold in the Civil War, and was captured by Commonwealth troops in 1644. When the victors demolished the building the condoms became trapped in a moist environment which preserved them for the next 342 years, until they were excavated by archaeologists in 1986.

Such condoms were certainly used in the 1700s and indeed the notorious lover Casanova had one. He referred to it as his 'English overcoat' in his *Memoirs,* whereas the English referred to them as 'French letters', a term that was still in common use in the middle of the last century. They were made of sheep gut and were meant to be reused. The condom is supposed to have got its name from an English army doctor, Colonel Quondam, who advocated its use among his men as a means of preventing venereal disease, but there is no reliable evidence that he ever existed. Condoms were openly advertised and sold in London in the 1690s by a Mrs Phillips and she is referred to in a comedy, *The Ladies' Visiting Day,* written by William Burnaby, in which Lady Lovetoy is accused of associating with Mrs Phillips.

Mrs Phillips and other purveyors of condoms sold their wares during the day in St James's Park, where the gentry were known to take their walks, and in the evening they peddled them around the various

29) If this had been dipped in vinegar it would certainly have been effective because sperm are disabled by acids.

theatre venues. These condoms came from a part of the intestines of a sheep that has a closed end and they needed to be kept moist to keep them flexible or at least soaked in water before use. They came with a pink ribbon at the base with which to secure them to the penis. Those who bought their condoms from the 'Cundum Warehouse' shop in St Martin's Lane were advised to wear two for safety because they tore easily.

The rubber condom appeared in the 1800s, following the discovery by Charles Goodyear that natural →latex could be made stronger and much more flexible if sulfur was added to it in the molten state. This resulted in all kinds of rubber goods becoming available and by the mid-1800s condoms were being manufactured. An advert even appeared in 1861 in The *New York Times* for 'Dr Power's French Preventatives.' The upright and uptight readers of this newspaper were appalled by such adverts and in 1873 the Comstock Law was passed which made it illegal to advertise any kind of birth control. While a similar law did not apply to residents of the British Empire, condoms were regarded by respectable people as unspeakably sordid and their distribution had to be by inconspicuous routes. Indeed the word 'condom' did not appear in the Oxford English Dictionary until 1972, such was the embarrassment it was thought likely to cause readers.

Victorian condoms were made of relatively thick rubber and were meant to be reused, but in 1919 an American, Frederick Killian, introduced a much thinner version made from natural latex and with a teat-end. By the mid-1930s more than 500 million a year were being manufactured in the US. They were dusted with chalk and some were even sold ready rolled but there were still many disadvantages associated with them. They could tear during use, they had a deadening effect on the man's sensitivity, they had a rubbery odour, they could not be lubricated with oils because these caused the rubber to weaken, they tended to deteriorate on storage, and some men were allergic to natural latex so they suffered intense discomfort following their use. A small step forward was the introduction of the *lubricated* condom by the Durex Company of the UK in 1957, which relied on a water-based lubricant, based on chemically modified cellulose. (Oil-based lubricants such as baby oil and hand creams should not be used with natural latex condoms.)

For those allergic to latex it continued to be possible to buy natural skin versions, and these are still available as Naturalamb condoms and

come from New Zealand. Most are used in America and Italy. Like the ones sold by Mrs Phillips 300 years ago, they are made from that part of a sheep's intestines known as the caecum, which is a sac-like compartment found near the opening of the large intestine. Supply of course is limited to one per slaughtered animal and even so the caecum must not come from an animal that is too young or too old because then it is either too thin or too thick. These skin condoms have the advantage of not gripping the penis along its length except at the base where there has to be an elastic band to hold it in place. They can be lubricated with oils.

The benefits of the latex condom and the natural skin condom were combined in an improved type of condom devised by research chemists and introduced in 1991: the polyurethane condom. Polyurethane is a →polymer which comes in many forms, some of which are ideal for insulation, some for car bumpers, some for disposable drinking glasses – and some for condoms. Polyurethane has been used to make body implants and surgical dressings that allow wounds to breathe without drying out, and without allergic reactions being reported. More than 12 million tonnes of polyurethanes are manufactured worldwide every year, although only a tiny fraction ends up in condoms. While polyurethane may not be as elastic as latex rubber, it is flexible enough for condom use involving normal penetrative sex. These condoms are transparent, super-sensitive, non-allergenic, unaffected by lubricants and an effective barrier against sperm and all sexually transmitted diseases.

Condoms need to be stretchable and polyurethane can be made with any degree of flexibility. Condoms were only the latest outlet for this remarkable polymer. In 1959 chemist, Joseph Shivers, who worked for DuPont in the US, developed an elastic version of polyurethane which they called Fiber K and launched as Spandex. It was ideal for corsets and girdles which were then popular items of female underwear. It became more famous in the 1980s as Lycra. It could be blended with other fibres to make stretchy textiles which were unaffected by perspiration, body lotions, and detergents, and in those respects it was much better than natural rubber. Soon Lycra garments were being worn as swim wear, ski wear, tights, dance wear and stretch jeans. The success of Lycra has not been without its problems – see box.

In For a Long Stretch

In 1989 five employees at the DuPont Lycra plant in Argentina tried to blackmail the company for $10 million by stealing top secret documents relating to the technological process by which the company's polyurethane elastomer was made. They flew from Argentina to Wilmington, Delaware, where DuPont is based, where they started negotiations with the company. From there they went to Milan, Italy, and thence to Switzerland, where it had been agreed they would hand over the documents to company representatives in return for a banker's cheque (regarded as unquestionably acceptable by all banks). They were to be given one that appeared genuine but was in fact bogus. However, the operation went wrong and the blackmailers refused to complete the deal, but they were later arrested in a car park in Geneva.

Polyurethane is the plastic of the bedroom and provides the foam filling for mattresses, the body-hugging materials of sex garments, and the protection of condoms. It is only this last use that concerns us here. Polyurethane condoms were launched under the trade name Avanti and made by the London International Group. (In fact Apex Medical Technologies of the US had earlier produced such a condom, which they called the Sensation condom, and got FDA approval in 1989 but it was never brought to market.) The Avanti condom is made from Duron, a particularly strong and flexible polyurethane. To begin with they were made ultrathin (0.04 mm) but then it was discovered that the breakage rate of 5% was unacceptably high, and so a slightly thicker (0.06 mm) version was introduced and the breakage rate fell to less than 1%. The Avanti is made to be wider than the penis-gripping latex condoms so that the user is less aware of it during intercourse. Tests showed that 80% of users preferred them to latex condoms due to their increased sensitivity.

Other companies have also produced condoms made from the polymer styrene-ethylene-butylene-styrene which has the advantage of being even stretchier than polyurethane, thereby reducing the risk of tearing during use. There are also condoms lubricated with a combination of silicone oil and nonoxynol-9 which acts as a spermicide. Nonoxynol-9 is also used as a surfactant in cosmetic products, but it has the ability to interfere with the acrosomal membrane which covers the head of a sperm, causing it to become paralysed, which is why nonoxynol-9 has been widely used in lubricants for condoms and especially for diaphragms. What it cannot provide is extra protection against diseases such as HIV.

Condoms have been attacked on religious grounds (because there are not ordained by God or his prophets) and moral grounds (because they are accused of encouraging promiscuity) and, for latex condoms, on health and safety grounds (the reason being that these were said to contain traces of nitrosamines, known cancer-causing agents). These nitrosamines are supposed to be formed from the phenylene diamine that is added partly to protect the latex against the damaging effects of oxidation and partly to increase its elasticity. The oddly named Chemical and Veterinary Investigation Institute in Stuttgart issued a press release in May 2004 claiming to have detected harmful levels of nitrosamine in 29 of 32 condoms they investigated. Although this news item is repeated on hundreds of sites on the web, the Chemical and Veterinary Investigations Institute itself remains somewhat elusive, having neither web site nor contact address, making one suspect that it might have been simply a press release designed to misinform. Its findings have yet to be published in a reputable scientific journal so it is impossible to judge their reliability. The German organization of condom makers naturally disputed these claims, pointing out that the research needed validating before it could be taken seriously and pointing out that studies by the University of Kiel, Germany, in 2001 could find no link between condom use and cancer.

We Just Got Carried Away Mum

How many mother's throughout history have heard, or dreaded hearing, those fateful words from their unmarried daughters? How many mothers, already burdened with too many children, were depressed to find themselves pregnant again? There is no excuse for a mature woman ending up pregnant as a result of taking chances, but for the young unpartnered woman it may be the unintended outcome of her first sexual experience. In earlier times it could threaten family disaster and end up ruining her chances of marriage and leading a normal life. She was a fallen woman.

Thankfully there was generally a wise-woman in every community who had the skill to perform an abortion, or to provide advice about natural chemicals that were known to cause a spontaneous abortion, such as the herb pennyroyal. Leaves of this were infused with hot water and the resulting drink would contain enough of the active agents, menthone and pulegone, to do the job. In unskilled hands there might be enough of these two deadly agents to kill. Failing access to a wise woman, the traditional remedy was to sit in a hot bath and drink half a bottle of gin.

Roman prostitutes were reputed to consume an artificial sweetener *sapa* because it both acted as a contraceptive as well as causing abortions, which it would do because these are symptoms of lead poisoning. *Sapa* was made by boiling sour wine in lead pans to give an intense sweet syrup that was in effect lead acetate. This was used in cooking and to preserve wine, and it was accused of causing all kinds of ailments including miscarriages. While *sapa* disappeared from history along with the Roman Empire there was another product of those times that persisted well into the 20th century and that was diachylon plasters, invented by Menecrates, physician to the Roman Emperor Tiberius (ruled 14–37 AD). These were designed to treat skin complaints, such as chilblains, chronic leg ulcers, and boils, and consisted of a paste of lead oxide and olive oil. The plasters led to an outbreak of lead poisoning among women in Birmingham, England, in the 1890s. They had discovered that if enough paste was scraped from diachylon plasters and eaten it would terminate an unwanted pregnancy.

Normal sexual intercourse is designed to make a woman pregnant and this happens when a sperm from the man merges with the woman's egg. Once an egg has been fertilized, it seeks to implant itself in the wall of the womb although this can take a few days, and it is those few days that offer an opportunity of preventing an unwanted pregnancy, either by taking a dose of the actual hormones that trigger menstruation or taking the drug mifepristone which achieves the same end. Hormone-based morning-after pills contain both the hormones oestrogen and progestogen. Four tablets have to be taken, two immediately and two 12 hours later, and each contains the oestrogen ethinyloestradiol[30] (50 micrograms) and the progestogen norgestrel[31] (250 micrograms). Together these will prevent a fertilized egg from implanting itself, but the side effect of these tablets can be a feeling of

30) Normally dispensed as a contraceptive pill under a variety of trade names.

31) Also known as levonorgestrel.

sickness and actual vomiting. If this occurs after the first tablets have been taken then a doctor will have to prescribe a further course of tablets along with something to stop the vomiting.

Another way to prevent a pregnancy is to take the drug mifepristone within three days of conception. Pharmacists in the UK can even supply it as an over-the-counter drug to women over the age of 16, who can give a reason for wanting it, and it has been available this way since the beginning of 2001. It was reported that 7% of sexually active women sought this treatment in 2005, generally giving condom failure as their excuse for needing it. Mifepristone was first made and tested in the early 1980s. It interferes with the progesterone that the body needs in order to maintain the lining of the womb in a state that will support a fertilized egg. Mifepristone binds to the receptors that need the progesterone and without the natural hormone they cannot make the proteins that the foetus needs, so the body rejects it. The drug comes as two 0.75 mg dose tablets, one to be taken immediately, the other 12 hours later (although some doctors recommend taking both tablets together). Tests in the late 1980s, on 400 women who thought they might have conceived and were given the drug, showed that none subsequently became pregnant.

In any event, there is no need nowadays for the indiscretions of youth to result in an unwanted baby and a blighted career-chemistry has seen to that. Even when chemicals are grossly misused so as to obtain sex forcibly by criminal means, it is still be possible to rectify the physical damage that rape causes, although the emotional damage will remain ... which brings us to the issue of date-rape drugs.

Issue: The Abuse of Chemistry for Sex Crime Purposes: Date Rape Drugs

Slipping someone a Mickey Finn in their drink in order to render them unconscious was a part of the plot of many film melodramas of the silent era, but it was based on fact. Mickey Finn was the landlord of the Lone Star Saloon on Whiskey Row in Chicago in the 1890s. His patrons often got more than they expected in the form of a few drops of chloral hydrate in their drink. They were then lured by one of Finn's female employees into a back room where they would soon become unconscious and be robbed of all their clothes and money before being dumped in the back alley.

The great German chemist Justus von Liebig discovered chloral hydrate in 1832 when he bubbled chlorine gas into ethanol (alcohol). The product was an oily substance that reacted with water to give chloral hydrate and this could be obtained from the solution as colourless crystals. These are very soluble in water and a few drops of that solution will induce a deep coma within minutes, which is why doctors prescribed chloral hydrate as a powerful sedative for patients who were in need of sleep. Of course these 'knock-out' drops appealed to those with criminal intentions, who could use them to drug a victim before robbing or raping them. (This was made a punishable offence in the UK as long ago as 1861 under the Offences Against the Person Act.)

Today other substances can be used and they are now referred to as date-rape drugs and there is widespread belief that this type of crime is quite common. It isn't. Clinical analysis of the urine or blood of women who claim to have fallen victim to their use, shows that very few have in fact been deliberately drugged. Most have passed out unconscious because they have drunk too much alcohol. In 2005, Michael Scott-Ham and Fiona Burton of the Forensic Science Service, London, reported in the *Journal of Clinical Forensic Medicine* the results of a 3-year study. They analysed 1014 cases of alleged date rape but found that only 21 of the victims (2%) had had their drink deliberately spiked. On this basis, of the 500 such cases reported annually in the UK, only 10 are actual date rapes. In the USA there are more than 250,000 reported rapes per year, but how many of these are drug assisted is not known, although on a comparable basis of 2% it would mean around 5,000. What was discovered in the UK survey was that while alcohol had caused most of the women to pass out, many of them had also taken recreational drugs such as cannabis, cocaine, or ecstasy. One woman in the report had traces of *four* such drugs in her blood.

The chemical agents used by those intent on rape are GHB, short for gamma-hydroxybutyric acid,[32)] which is a powerful muscle relaxant, ketamine,[33)] which is used as an anaesthetic, and flunitrazepam,[34)] which is better known by its trade name Rohypnol and is a powerful antidepressant. They have a variety of street names, such as liquid X, liquid E, and cherry meth for GHB, kit-kat and special K for ketamine, while Rohypnol pills are referred to as roofies, roachies, rope, and roche[35)] as well as more descriptive names such as circles, forget me pills, and lunch money. These three chemicals will cause a person to become dizzy and then completely unconscious, so much so that when they finally come round many hours later they may find it impossible to recall anything that has happened to them. This amnesia seems to be a permanent loss of memory although some of the victims have flashbacks in the days following the rape.

GHB occurs naturally in the body in small amounts, and in the 1980s it was sold as a nutritional supplement and used particularly by bodybuilders to enable them to promote muscle growth. However, in some people it caused seizures and coma and it was banned in the US in 2000. Nevertheless it is easily made and widely used to create a feeling of euphoria, or to reduce the withdrawal symptoms of other recreational drugs.

32) Also more correctly known as 4-hydroxybutanoic acid; it has the chemical formula $C_4H_8O_3$.

33) Ketamine has the chemical formula $C_{13}H_{16}ClNO$.

34) Rohypnol has the chemical formula $C_{16}H_{12}FN_3O_3$,

35) It is manufactured by the pharmaceutical company Roche.

If a woman suspects that she has been deliberately drugged and raped then forensic analysis of her blood and urine should be carried out as soon as possible after the suspected offence because some of the drugs, and especially GHB, are rapidly lost from the body. She should report immediately to the police so that a urine sample can be taken. The various date-rape drugs can be detected at concentrations down to nanograms per millilitre (parts per trillion) which means that even after a week it may still be possible to detect some of them, such as Rohypnol. While this drug features on many internet sites warning of date-rape, it was not detected in any of the 1014 women covered in the Scott-Ham survey. The reason for this is undoubtedly due to the inclusion since 1999 of a marker chemical in the drug and this turns any drink to which it is added a bright blue colour.

While Rohypnol is legal in Europe and countries like Mexico, it is illegal in the US, where it is listed as a Schedule III drug, in other words it is currently accepted for medical use and can be abused but it is not addictive.[36] The fact that it was illegal probably added to its attraction among the young people in that country in the 1990s, as evidence by police raids in some southern states when tens of thousands of Rohypnol tablets were seized.

There are always those would seek to misuse chemicals for their own evil ends but fear of date-rape drugs is far more widespread than actual cases of drugging. If you are worried about this happening to you then follow some simple rules: only drink alcohol in the company of friends; don't drink on an empty stomach; and never accept a drink from a stranger. If you do pass out, and come round hours later in some strange place, and suspect you have been raped, try to get help immediately and resist the urge to urinate until a sample of urine can be collected under proper supervision. That way you might help convict the rapist who has abused you and hopefully see him jailed for a long time.

36) Schedule I drugs are not accepted for medical use and are highly addictive, while schedule II drugs are accepted for medical use but are known to be addictive.

4
Better Living (II): Improving Our Diet

A small arrow printed before a word in the main text indicates that there is more information on that topic in the Glossary.

News from the Future

Global Times News, 21 March 2025

Yet More Land Goes Back to Nature

Thanks to the increased yield of ammonium nitrate from self-sustaining NitroFix units another million hectares of farmland will revert to the wild this year.

In 2010 the United Nations reported that low-yield organic farming was damaging the Earth's ecology by requiring more and more tropical forest to be cleared to grow such crops. Not only did such farming produce falling yields as the nitrogen in the soil was used up, but the crops so produced were then transported thousands of miles to supermarkets in Europe and the USA.

The UN report boosted the demand for ammonium nitrate and that in its turn led to the development of small scale nitrogen fixation units making ammonium nitrate and fuelled by electricity generated by windmills or water mills. NitroFix units now supply much of the fertilizer nitrogen the world needs with almost no adverse environmental impact. Not only that, but it has put an end to fertilizing food crops with human sewage which is now only used to fertilize woodland, energy crops, or for methane generators.

"It's a win-win situation" said the Secretary of the United Nations as she opened the ten thousandth NitroFix unit in Kenya, and then went on to inspected the newly enlarged national nature reserve.

Page 8: Organic food prevents cancer claims leading compost distributor.

During the past one hundred years chemists have been able to increase world food production in two ways: through the production of fertilizers which have doubled and tripled crop yields, and by creating better pesticides so that insects and plant diseases take less of them. Even though this century will see significant changes in agriculture, with the introduction of genetically modified plants, there will still be the need to replenish the soil, especially with nitrogen, and to protect

Better Looking, Better Living, Better Loving. John Emsley

ISBN 978-3-527-31863-6

crops. Organic farming does neither very well and if we were to try and feed all of mankind this way then it would be a planetary disaster, because we would have to cultivate every hectare of soil that is available. A news item from the future like the one above will one day have to happen if this planet is to be sustainable and millions are not to starve.

Meanwhile in the West we continue to agonise over our food while we continue to consume more and more, and yet we get plenty of advice about what we should and should not eat. A lot of this advice is quite sensible when it comes from a properly qualified dietitian but some of it is simply supposition and some may be just alarmist speculation. Food packaging also carries messages such as 'contains no added sugar', 'low salt', 'free from artificial colours, flavours, and preservatives'. Of course if you go into a supermarket you can still buy large amounts of two of these apparently dangerous foods in the form of bags of sugar and packets of salt. When it come to avoiding 'artificial' ingredients you might think that only *natural* food additives, such as stock cubes, spices, herbs and soy sauce, are safe and quite unlike the *chemical* additives that you worry about. The impression is often given that food manufacturers are sneaking all kinds of things into our food to the benefit of their profits and to the detriment of our health. Maybe some are, but most responsible food companies employ food chemists and analysts and these scientists are there to make sure that what they produce is not only safe to eat but will still be safe to eat at any time up to the date stamped on the container, and that it fulfils a customer's expectations as to how it looks, smells, tastes, and nourishes.

Some who campaign on food issues spread their dietary advice with almost religious zeal and they are skilled in giving their views the right spin to guarantee they will be widely reported. Take their advice with a pinch of salt. We are also exposed to many other messages about food, either from celebrity chefs on how to prepare their favourite dishes, or from the authors of diet book telling us how to lose weight when we've eaten too much. Because food is something that we enjoy several times a day there will always be an audience for advice about the things to eat.

This chapter is different because it aims to look a little deeper at food and by deeper I mean at the molecular level. Here you will learn that carbohydrates are *essential* components of a healthy diet; that one particular carbohydrate may save millions of lives; that that by adding something to common salt the lives of many around the world is being changed for the better; how some functional foods might make us healthier by boosting and feeding the microbes that live inside us; and why spicy food can seem so hot even when cold. We'll also take a brief look at mother's milk, royal jelly, and so-called functional foods.

Carbohydrates

The term →carbohydrate covers a large number of chemical compounds. The popular name is carbs, but this has become a somewhat judgemental term which assumes they are something to be avoided. You probably already know that there are simple carbohydrates, such as glucose and fructose, and complex carbohydrates, such as starch, glycogen, and fibre. The scientific name for ordinary sugar is sucrose, and this consists of a glucose and a fructose molecule bonded together. Starch on the other hand consists of hundreds and thousands of glucose molecules linked in chains, and the same is true of fibre, which is mainly cellulose. Starch and cellulose differ in the way the glucose units are joined together. We have enzymes in our body that can break down starch into its glucose units and thereby digest it, but we cannot break down cellulose, which may not be a bad thing because as fibre it helps our food move easily through our intestines.

Just how important carbohydrate is can be seen by examining royal jelly, a complete food that provides all the nutrients that are needed for life – at least for bees. Royal jelly is secreted by worker honeybees to feed young larvae and the queen bee herself, but whereas the young larvae only get it for the first three days of their lives, the queen lives on it for all her life, which might exceed five years. (Most other bees, the workers and the drones, live only a few months, and they feed on pollen and honey.) Some people believe that royal jelly has ingredients that increase fertility and longevity and take it for these reasons. Analysis of royal jelly shows it to be a mixture of carbohydrates, proteins, fats, vitamins and minerals – all the expected ingredients that are needed for a healthy human life as well. It is particularly rich in vitamins B_1, B_2, B_3, B_6 and B_{12}. The composition of royal jelly is 65% water, 18% carbohydrate, 12% proteins, 4% fats, and 1% minerals and vitamins. It is slightly acidic, with a →pH of around 4, and it is slightly antiseptic. Whether it really does contain the secret molecules that guarantee a longer life has yet to be revealed, but what we should note is that among its nutrients, carbohydrates are the most abundant.

If you ate nothing but carbohydrate you would die. If you ate *no* carbohydrate at all you might well end up the same way. That appears to contradict the prevailing view that all carbs are a threat to health, primarily because excess carbohydrate causes obesity. In 1991 it was said that 12% of Americans were obese, a figure that increased to 18% by 1998, and to 26% in 2004. Where the Americans lead, the rest of the world tends to follow. After we have fallen into the sin of gluttony and been punished with obesity, then we are offered a road to salvation via a carb-free diet. That advice will certainly go a long way to solving the problem – as it did in the 1950s and 1960s when the recommended way to lose weight was to cut out sugar, bread, pasta, and potatoes, all high in carbohydrates. In the 1970s and 1980s sugar was fingered as particularly dangerous, with books like the late John Yodkin's *Pure, White and Deadly* driving the point home with almost fanatical fervor. The late Dr Atkins' diet of the 1990s also preached the message of no carbs. The GI (short for glycemic index) diet of the 2000s does not go quite so far but it restricts them to the least damaging ones. The glycemic index grades foods according to the effect they have on the glucose levels of the blood. Those that release carbohydrate at a controlled rate and so keep the glucose levels more stable have a low GI rating and are to be preferred. Those which are rapidly digested and absorbed, thus leading to marked fluctuations in blood sugar levels, have high GI ratings and are to be avoided.

Ultra low-carb diets are wrong and go against the advice of dietitians who say that we should aim to take in around 60% of our nutrients as carbohydrate, 20% as protein, and 20% as fat. (The nutrients in royal jelly are roughly in line when it comes to carbohydrate, but it has more protein and less fat. Its composition is 55% carbohydrate, 35% protein, 10% fat.) There really is no such thing as a bad carb, but there can be a bad diet that contains far too many of one sort, of which sugar is the most common and the one we need to eat less of. (In 2000 the average American consumed around 75 kg of refined sugar annually, which alone provides 30% of the average person's food calorie needs.)

If you exclude *all* carbs from your diet, you will lose weight quickly and easily – but is this the best thing to do? Knowing more about the biochemistry of carbohydrates might well help you make a decision that is right for you. One organ in our body needs a constant supply of carbohydrate in the form of glucose, and requires about 140 g of this per day, and that is the brain. If our diet does not provide it with the glucose it needs then our metabolism has to manufacture substitute chemicals, known as ketones, which can be used instead. Other parts of the human body also need carbohydrate such as red blood cells and our kidneys. If there is none available then the body will even make carbohydrate from proteins.

Carbohydrates come in all shapes and sizes, and there are lots of them. All life on this planet ultimately rests on the ability of plants to take carbon dioxide out of the air and convert it to carbohydrate. (Oxygen gas is released in the process.) Perhaps not surprisingly, staple foods for humans around the world are carbohydrate-rich crops, such as rice, potatoes, maize, corn, flour, and chickpeas. It is possible to exist on one vegetable alone: potatoes. This was proved in the 1920s when a 25-year-old man and a 28-year-old woman lived entirely on potatoes and nothing else for six months and were found to be in good health.[37] Nor were they obese. The carbohydrate content of potatoes is of the order of 80% or more. In some parts of the world, such as China and Japan, the concept of a meal without rice is almost unthinkable; indeed the Chinese use the same word for rice and for food. Nor is a carbohydrate-rich diet necessarily a recipe for poor health, witness the longevity of the Japanese whose average lifespan is the longest of any group of people.

Glucose and fructose are the most abundant carbohydrates in Nature. Joined together they form sucrose, better known as sugar. Variety can come from just one of these simple carbohydrates. Thus two glucose molecules can link in several different ways to form either trehalose, maltose, isomaltose, cellobiose, or gentiobiose. The only difference between the first two of these is that the former has its two glucose molecules facing the same way, whereas in trehalose one is upside down with respect to the other. In the other double-glucose molecules the two components are joined together at different points in their structure.

One particularly important complex carbohydrate is glycogen, which the body makes in order to store glucose, and it consists of hundreds of glucose units in a chain with other glucose units attached as

37) Such a diet could not provide all that the body needs on a long-term basis.

side chains and all curled into a space-saving spiral shape, making storage within cells easier. In addition to glycogen there are other carbohydrates that are an essential part of human metabolism. Can we make these within the body, or have they to be taken in as part of our diet? The answer appears to be that we can make them, and make them from whatever food we eat, although there is evidence that some carbohydrates, such as mannose, can be utilised more efficiently if they are absorbed from our food. This was shown by giving volunteers radioactively labelled mannose and finding that this was used to make components needed for the liver and intestines.

Some carbohydrates are essential for cell-to-cell communication and we depend on carbohydrate attached to cells to confirm that the cell is healthy or signal that something is wrong. Eight different carbohydrates are involved in cellular recognition and they are responsible for the different blood types, and that difference may be as small as just one simple carbohydrate. For example Type B blood has a galactose molecule whereas in Type A this is slightly modified with a nitrogen atom attached. These may not be big differences chemically speaking but they are enough to threaten the life of someone who is given a transfusion of the wrong blood. As we saw in Chapter 2, it is the carbohydrates on the outside of a cell wall which viruses exploit in order to invade it. Equally important are the carbohydrates which trigger off the production of antibodies to attack the invading viruses. Never let it be said that carbs are something we can do without.

We digest carbohydrate at different rates at different sites along the human gut, beginning in the mouth and ending in the colon. If you chew a piece of bread and then keep it in your mouth for a while it begins to taste sweet. This is because the enzymes in our saliva are able to break down the starch it contains into its glucose components. Some carbohydrates resist digestion and these we call fibre and they consist mainly of cellulose. Some carbohydrate we could digest but have difficulty doing so because it has been altered in cooking and turned into what is referred to as resistant starch. This is formed when we re-heat starchy foods that are moist, such as mashed potatoes and pasta. The molecules of starch then stick to one another in a way that prevents digestive enzymes getting at them.

The large bowel, also known as the large intestine or the colon, is where there is a rich culture of bacteria waiting to feast on things like complex carbohydrates, which they break down into the short-chain

→carboxylic acids. These are: acetic acid, which can be turned into energy by reaction with oxygen; propionic acid, which the body simple excretes in the urine because it has no use for it; and butyric acid, which is taken up by the cells lining the colon and there used to regulate their growth and maybe protect them against cancer. The table below summarises the main types of carbohydrate in our food and the way our body deals with them.

Carbohydrates in the Diet and in Our Gut

Type	**Chemical names**	**Foods rich in**	**Where digested**	**Speed of digestion**
Simple sugars	Sucrose, glucose, fructose, etc.	Sweets, cakes, biscuits, soft drinks, fruits, honey, jam.	Stomach	Very fast
Starches	Amylose, amylopectin	Cereals, rice, potatoes, pasta.	Upper intestine	Slower
Fibre	Cellulose, pectin	Wholegrain cereals, fruit, bran, beans.	Not digested	Nil
Resistant starch	Amylose – see text	Reheated foods such as potatoes and rice; cornflakes.	Partly digested	Very slow, if at all

Starch is a component of seeds, tubers and roots. It has two major components: amylose and amylopectin. Amylose is a polymer consisting of linked glucose molecules and a typical strand contains around 2,500 glucose molecules. Amylopectin has even longer chains with lots of shorter carbohydrate chains attached to it, and this type of starch makes a good thickening agent and is ideal for gravy and sauces. Starch can be modified in several ways, by heating it or treating it with various chemicals, and as such it is added to products like yoghurts to produce a creamier texture in the mouth. And when low fat foods are the order of the day, such as 'light' mayonnaise and yogurts, then modified starch can duplicate the creamy texture without the need for oil or cream – see box.

A New Kind of Starch

Collaboration between food chemists working for the Dutch company Avebe, which produces starch from potatoes, and the Netherlands Applied Scientific Research Organization, (better known as TNO, which is short for **T**oegepast **N**atuurwetenschappelijk **O**nderzoek), have found a way to modify starch so that it has the useful jelly-like characteristics of gelatin, in other words their starch becomes liquid when heated and sets solid on cooling. The food chemists found an enzyme in the heat-tolerant bacterium *Thermus thermophilus* that was able to link amylose and amylopectin together to produce the desired result. The product has a clean flavour like that of potato starch and it is destined for products such as yoghurts, spreads, puddings, and imitation cheeses. There are also non-food applications for the new starch, as adhesives and in photographic films.

To maintain a constant level of glucose in the blood the liver either converts it to make glycogen if there is an excess, or releases it from its glycogen store if there is too little. Those who engage in strenuous exercise use up the energy in their body and that means tapping its glycogen store. There can be up to 1 kg of available glycogen, capable of yielding as much as 4000 food calories of energy. When the glycogen has been used up then muscles will cry out for rest and that is likely to happen when a person has engaged in strenuous exercise for an hour or so. If no rest is forthcoming then the body begins to generate energy in other ways. Alternatively we can supply glucose by drinking a sugary drink of the kind designed for this purpose.

A marathon runner who completes the 25 mile (40 km) course in 3 hours will consume as much energy as a 1 kW electric heater uses in an hour. Although he or she generates a lot of heat in the process, the body stops itself from overheating by evaporating about 2 litres of sweat from the skin. Along with the sweat, the body also loses electrolytes, namely up to four grams of sodium and about half a gram of potassium. Clearly the body's immediate requirement is for water, but other things can be added to this which might well help. A typical sports drink also contains glucose, which the body can absorb directly into the blood stream from the stomach, but levels in the drink must not be too high because that delays absorption of the much needed water.[38] A typical sports drink has around 50 g of glucose per litre. The

38) If there is too much sugar in a drink then this has to be diluted in the stomach by taking water from the body before it can be absorbed.

amount of sodium in it will be around half a gram per litre (equivalent to 1.25 g of salt per litre) and there may be other things as well.

Not surprisingly when an athlete consumes a sports drink it boosts his or her performance more than if they simply refresh themselves with a drink of water. The Japanese were ahead of the game and had already introduced such a health drink in the 1980s under the not inappropriate name of *Pocari Sweat* whose sales were eventually to exceed 200 million bottles. Similar drinks are Gatorade in the USA[39] and Lucozade Sport in the UK. Gatorade contains 56 g of carbohydrate, 4.5 g of sodium, and 120 mg of potassium per litre, while Lucozade Sport contains 64 g of carbohydrate, 1 g of sodium, and 100 mg of potassium per litre.

One natural carbohydrate has recently come to prominence because it promises quite unique benefits and yet it is mentioned in the Old Testament of *The Bible.*

Manna from Heaven ...

... came in the form of the carbohydrate trehalose and, as mentioned above, this molecule consists of two glucoses joined together. It is a naturally occurring, sweet-tasting carbohydrate, widespread in Nature and a source of energy for many living things, including bacteria, fungi, insects, plants and invertebrates. When eaten, trehalose is digested in the small intestine where it is broken down into its constituent glucose molecules by the enzyme *trehalase.* It is only half as sweet as sugar, but its taste persists for longer. Foods such as honey, bread, beer, and mushrooms, contain trehalose but only in tiny amounts, although shiitake mushrooms may contain up to 20%. Trehalose is also present in the poisonous mushroom fly agaric (*Amanita muscaria*).

Microbes such as bacteria and yeasts readily produce trehalose but only a few plants make it, yet for those which do it is literally a lifesaver because it enables them to survive even the most prolonged drought. When it rains they miraculously come back to life. These so-called resurrection plants use this carbohydrate to preserve essential

39) Named after the University of Florida's football team, the Gators, who helped test the drink in the 1960s.

parts of their cellular structure when almost all water has gone. Some plants can lose over 95% of their water content and still survive. Trehalose also protects living cells against the stresses of heat and cold.

Humans have known of trehalose since Biblical times and it appears to have been the manna that God provided for the Children of Israel as they wandered for 40 years in the Sinai wilderness before reaching the Promised Land.

> *The Lord said unto Moses, I will rain bread from heaven for you.*
> *Each day the people shall go out and gather a day's supply.*
>
> [Exodus, 16:4]

> *Israel called the food manna; it was white, like coriander seed,*
> *and it tasted like a wafer made with honey.*
>
> [Exodus, 16:31]

Everyone was commanded to gather an 'omer' of manna, a quantity that we can no longer recognise, but it was possible to collect this amount relatively easily.[40] All we can really deduce from the Biblical account is that manna came in small pieces and was sweet to taste. According to the *Encyclopaedia Britannica,* the name was derived from the Hebrew phrase "*man hu?*" which translates as "what is it?" and that might well have been what the Israelites said when they came across large numbers of small white pellets, which they discovered were sweet-tasting. Nor did God provide manna only for the Jews; the Bedouin tribes of the region continued to gather manna as late as the 20th century, treating it as a delicacy.

What exactly was manna? Its sweetness, and the fact that it was available in an arid land, suggests it was rich in trehalose. This being so, then there are several possible sources. It may have been the cocoon of a parasitic beetle, called trehala mana, which contains around 25% trehalose. A possible contender is the solidified juice of the flowering ash, also known as the manna ash (*Fraxinus ornus)* which oozes from its bark and solidifies. This too is collected and sold commercially. Another suggestion is that manna is the lichen *Lecanora* which curls up into balls when there is a drought. These can be blown by the wind and are sometimes collected and used to make

40) The Revised English Bible (Oxford University Press, 1989) says that an 'omer' was about 4.5 litres.

bread and jelly. There is even the possibility that manna was the flaky material that aphids excrete and which is known as honeydew. It accumulates on the leaves of trees and is the main source of food of mosquitoes. It too has been eaten by humans on account of its sweet taste.

In the 1980s, John Crowe of the University of California at Davis was the first to suggest that trehalose was the secret of how plants survive in times of extreme drought. In 2002 a group of researchers at Cornell University in the US genetically engineered *Indica* rice to synthesise trehalose. (*Indica* accounts for 90% of rice grown worldwide and includes basmati rice.) They inserted into the rice the two enzymes that *E. coli* uses to make trehalose. Thomas Owen who led the group hopes that the same might be done to other crops like maize, wheat, and millet to make them more resistant to drought and heat stress and make it possible for them to grow in more arid regions.

Trehalose is chemically very stable. It dissolves in water to the extent of 700 g per litre at room temperature, and the solution can be boiled without any decomposition; trehalose can even be heated at 120 °C for more than an hour without undergoing any change. Trehalose has a special relationship with water, attaching two water molecules to itself and clinging so tenaciously to them that they are almost impossible to remove. The cluster which forms is extremely strong and rigid and it is this which enables it to protect biological structures against water loss and high temperatures.

Trehalose has attracted the attention of food manufacturers because it stabilises other ingredients, such as starch, fats, and proteins, although its sweetness can mask unpleasant tastes and odours, such as those that might arise when fats begin to react with the oxygen of the air and go rancid. Trehalose preserves the flavour of dried foods and indeed the infamous dried egg of World War II could again become acceptable, because a little trehalose enables it to retain the taste of fresh egg when it is reconstituted. Trehalose is being used to preserve foods by treating them with a solution of it and then air-drying them, this being a cheaper alternative to vacuum-drying. Moreover, apple slices that are dipped in such a solution before drying do not go brown, and other fruit and vegetables treated with it retain their colour, fresh flavour and texture. Trehalose improves the quality of sweets such as caramel, toffee, chocolate, and chewing gum.

Today, trehalose is produced in large amounts by an enzyme process patented by Takanobu Hayashibara of the Amase Institute at

Okayama, Japan. He reported his new enzyme system in 2002, which he had extracted from a bacterium. This will convert starch to trehalose in high yields, bringing down its cost to 1% of what it had been when it was extracted from natural sources such as yeast. Hayashibara also reported that trehalose suppresses human body odour, especially that given off by old people who produce the somewhat odorous chemicals 2-nonenal and 2-octenal[41] in their skin. When they use a 2% solution of trehalose as a body lotion it reduces the emission of these smelly compounds by about 70%. Maybe one day it will find cosmetic uses and no doubt be added to deodorants and body lotions, as the following advert from the future shows:

The Senior Citizen Courier 2025
[Advertisement]

Golden Halo

Sensational New Body Lotion!

Worried that you may be giving off 'old folks' body odor? Do your grandchildren turn up their noses when you ask them to sit on your knee? Then try **Golden Halo**, the new body wash and lotion that guarantees you will smell 'shower-fresh' all day. **Golden Halo** contains the miracle ingredient *trehalose* that actively suppresses those tell-tale skin odors that signal old age.

Be nice to be near!
Buy Golden Halo today!

Just as molten sugar can solidify into a transparent material (and is used to make the fake window glass that actors can crash through without cutting themselves) so trehalose can form a glasslike solid and this has been suggested as a way of preserving antibodies. In 1985 Dr Bruce Roser,[42] working for the Quadrant Research Foundation in Cambridge, England, added trehalose to a solution of antibodies and then allowed the water to evaporate at 37 °C. When these were rehydrated they were found to have lost none of their activity in contrast to antibodies dried without trehalose which were inactive. Even after years of storage at room temperature, trehalose-dried antibodies work well. Enzymes can be similarly protected, and so can vaccines.

In response to an approach from the World Health Organisation (WHO), Roser has developed a trehalose-based technology for vaccines which means they can be stored without the need for refrigera-

41) Chemical formulae are $C_8H_{15}CHO$ and $C_7H_{13}CHO$ respectively, each with a double bond next to the aldehyde (CHO) end-group.

42) Roser is now (2005) Chief Scientific Advisor to Cambridge Biostability.

tion. This would allow most of those children who are not currently vaccinated to be so treated at a cost their impoverished countries could afford. More than half the vaccines produced in the world are not used because they are damaged by changes in temperature. They need to be kept refrigerated in hot climes and this is a major part of their cost and why so much is wasted. A vaccine will be discarded if a nurse or doctor suspects it has not been kept cold. Of the 130 million children born each year, around 100 million are vaccinated, but of the 30 million unvaccinated children more than 1 million die from preventable diseases.

Vaccines are mixed with trehalose solution and spray-dried to form microscopic glass-like spheres which are then suspended in an inert liquid. As such they are stable and the vaccine is only reactivated when the liquid is injected into the body where water dissolves out the trehalose and releases the vaccine. Roser's method allows mixtures of vaccine powders to be stored without risk of their reacting with one another, and they can even be injected together.

Low I_2 Means Low IQ

Or, to put it more positively, by adding small amounts of iodide to common salt we can prevent the medical condition known technically as cretinism, a life-long affliction of innately low intelligence caused by iodine deficiency, and which causes permanent brain damage. Nor does this simple dietary requirement cost very much. To provide all 6.5 billion people on the planet with the 70 micrograms of iodine they need every day would require only 166 tonnes of iodine per year, which represent a mere 2% of the iodine the world produces annually. This is an amount that chemists could easily extract from seaweed, making it a sustainable resource, and indeed this was once the way they obtained this remarkable element.

Iodine and iodide are chemical terms, used interchangeably when talking about the dietary requirement of this element. However, they are different. Iodine is the chemical element and exists as the molecule I_2 and this is how it was used as an antiseptic. It is not how it is found in Nature. There it is present as *iodide*, which is a negatively charged iodine atom, I^-, and this must occur in conjunction with a positively charged atom such as potassium (K^+). When iodide gets into the human body it undergoes a series of chemical reactions that transform it into the molecules that the body needs.

As long ago as 1960 the World Health Organization (WHO) undertook a global review of the disease goitre which, with its grossly swollen neck, is a well-recognized symptom of iodine deficiency. Iodine deficiency mostly affects school-age children but they don't exhibit the obvious symptom. We now know that goitre is not the best indicator of iodine deficiency because, once established in an individual, it responds only slowly to changes in the iodine content of the diet. Today iodine is assessed by the analysis of urine samples and when levels of excretion fall below 50 micrograms per litre it signifies a lack of this element.

Some degree of iodine dietary deficiency was estimated to affect 750 million people in the developing world in 1990 with around 10 million were suffering from stunted growth and mental retardation. In 1993 the WHO published a database of iodine deficiency based on the number of goitre victims in 121 countries where the element was in short supply. This shortage was most prevalent where soils have suffered repeated glaciations or heavy rainfall which had removed most of the iodide that may once have been present, because this is a particularly soluble material and so easily washed away. Those most at risk lived in India and China.

The WHO urged 110 national governments to make iodized salt mandatory. By 2004, 56 of these countries had responded positively and most of the remainder were on their way to complying, but not all, and 14 were still failing to supply their people with this simply remedy. In addition the United Nations General Assembly held a Special Session for Children in 2002 when it adopted a plan to maintain a databank on iodine deficiency and to assist member countries in ensuring that their populations had access to iodized salt. This they could do by working closely with various agencies, such as the International Council for Iodine Deficiency Disorders.

Iodine deficiency has its most devastating effects on babies while they are in the womb and during their first three months of life. These are the critical periods of brain development and lack of iodine leads to irreversible effects and permanent mental retardation and on average such children have an IQ of around 85 compared to the IQ of 100 of children who develop normally. Many exhibit the symptoms of brain damage due to lack of iodine, and this condition affects as many as one child in seven in some parts of the world. A woman needs double the normal iodine intake during pregnancy and when breast feeding if her baby's brain and nervous system are to develop properly, and the suggested amount is 140 micrograms a day.

Various ways to treated iodine deficiency have been used, including iodine solution as a medicine to be taken twice daily. This medicine is a mixture of 5 g of iodine and 10 g of potassium iodide in 100 ml of water, the standard dose being 1 ml. The iodine is quickly transported to the thyroid where it can be stored before being turned into the iodo-molecules that are the hormones the body needs. A better way of administering this form of iodine was to dissolve it in an oil and give it as intramuscular injections and this was done with much success in Papua New Guinea, and then in China, Latin America, and Africa. The benefit of iodine had been demonstrated in New Guinea in 1967 with the help of 8000 women, half of whom were given injections of iodised oils. Four years later, when around a thousand of them had given birth to babies it was found that among those who had received the iodine only 7 babies were born who were brain damaged due to lack of iodine, whereas among the others there were 26 so affected. This method of treatment is now only used in areas of acute iodine shortage; for the rest of the world the best way to ensure the necessary minimum intake has been to add it to common salt, of which most people consume between 5 and 10 g a day. The iodine in salt is added as either potassium iodide (KI) or potassium iodate (KIO_3) at between 20 and 40 ppm, depending on whether the climate is temperate (when potassium iodide is used) or tropical (when potassium iodate is used). Potassium iodate is more stable than potassium iodide in hot and humid climates. Ideally everyone in the world should consume iodized salt in which the level is at least 15 ppm.

The average person has between 10 and 20 mg of iodine in their body, most of which is in the thyroid gland, where it is present in two hormone molecules, thyroxine[43] and liothyronine,[44] having four and three iodine atoms respectively. These regulate several metabolic functions, in particular the body's temperature, so are needed throughout life. Iodine is also needed for normal growth and development. If we lack iodine this will eventually manifest itself as goitre and

43) Chemical formula $C_{15}H_{11}I_4NO_4$.
44) Chemical formula $C_{15}H_{12}I_3NO_4$.

hypothyroidism, which means a person continually feels listless and cold.[45)]

People living in developed societies get the small amount of iodine they require from their diet. Milk is a major source and its level has increased in recent years due to its supplementation in cattle feed and the use of iodine antiseptics to sterilise udders before milking, although this has now given way to the practice of sterilising *after* milking, cutting off a useful, if accidental, addition of iodine to milk. Foods naturally richest in iodine are mackerel, cod, and particularly haddock, a typical 150 g (6 ounce) portion of which contains 300 micrograms. Other foods rich in iodine are yoghurt, sunflower seeds, and mushrooms. The vegetables which take up most iodide from the soil are cabbages and onions which can have 10 ppm (dry weight). However, some foods, such as cassava, maize, bamboo shoots, and sweet potatoes can interfere with iodine uptake by the body, and they particularly threaten those living in regions where dietary iodide is already in short supply.

The linking of goitre to iodine deficiency was known almost 200 years ago but early attempts to alleviate the condition met with limited success. A French doctor, Jean-Francois Coindet (1774–1834), introduced iodine into medicine in the form of a solution of iodine and potassium iodide dissolved in alcohol, and he gave it to patients with goitre. That was in 1820. He knew of an earlier treatment for this disease which recommended eating seaweed ash and he suggested that since this was a rich source of iodine, it might well be that this was the active ingredient. He was right. Unfortunately when patients with goitre were treated with Coindet's tincture of iodine they suffered such severe stomach pains due to its irritant effects that the treatment was abandoned. Nevertheless Coindet's tincture of iodine became the accepted treatment for wounds for more than a century, although it has now been replaced by much less painful chemicals.

The need of the thyroid gland for iodine was finally proved in 1895 by a Dr Bauman who spilled some concentrated nitric acid on a sample of thyroid tissue and saw purple fumes of iodine rising from it.[46)] Finally in 1916 an American biologist, David Marine, of Ohio, proved

45) It is also possible to have a thyroid malfunction of too much activity known as hyperthyroidism with symptoms of restlessness and hyperactivity

46) The name iodine comes from the Greek word iodes meaning violet.

that goitre could be cured with iodine and, better still, that it could be prevented if iodide was taken as a dietary supplement. He suggested that the best way to do this was to add it to common salt. Iodized salt was first introduced in the USA and Switzerland in the 1920s, where it successfully eliminated goitre from the population. Sadly it was to take another 70 years before this simple remedy was available in places where it was most needed.

World-wide industrial production of iodine is about 13,000 tonnes per year, mainly coming from Chile and Japan. Known reserves of easily accessible iodine amount to around two million tonnes worldwide. From the 1820s to the 1950s, iodine was extracted from dried seaweed and this could once again be a source of the mineral in a self-sustaining future. Kelp contains 0.45% iodide (dry weight) and its ash is 1.5% iodide, permitting 15 kg of the element to be extracted from a tonne of seaweed ash. Seawater has only 0.06 ppm of iodide and yet kelp concentrates it. Soils on the other hand have an average 3 ppm of iodide, and in some regions, such as the Baraba Steppe of Russia, iodine in soils can be as high as 300 ppm, and soils along the coasts in Japan and Wales can have 150 ppm.

In Nature there is an iodine 'cycle' and indeed thousands of tonnes of iodine escape from the oceans every year as iodide in sea spray or as molecules containing iodine, which are produced by marine organisms. Marine algae emit volatile iodine as iodomethane (CH_3I) and diiodomethane (CH_2I_2) and these may even moderate the world's climate by helping cloud formation. Some of this iodine is deposited on land where it may become part of the bio-cycle. Recently it has been shown that rice plants also emit iodomethane and that this accounts for about 4% of that which is present in the atmosphere.

Naturally occurring iodine consists of only one isotope, iodine-127, which is not radioactive. However, the iodine produced in nuclear reactors is a heavier isotope, iodine-131, which is dangerously radioactive. This was released in large amounts following the nuclear accident at Chernobyl, Russia, in 1986. Iodine-131 has a half-life of eight days but because iodine can be absorbed rapidly into the human food chain, by getting into the milk of grazing animals, or directly into the water supply, it presents a real danger. To counter its effects, potassium iodide tablets were issued to those most at risk and these limited the amount of radioactive iodine that was absorbed, thereby limiting the damage it could inflict.

Without a doubt we could make goitre and cretinism a thing of the past for all people, and all this requires is for salt to be iodized in those parts of the world where there is a risk of iodine deficiency. Nor are we talking only of developing countries, because there is now a growing concern that iodized salt is not being used in developed countries. In the UK, for example, only 2% of the salt sold in supermarkets is iodized, and there are parts of that country where women now take in so little iodine that they could be putting their unborn babies at risk.

Despite what is commonly believed, a little salt is good for us, although clearly if we are advised by our doctor to cut down on it we must, because too much can be harmful to those with high blood pressure and heart disease. (People who continually eat highly salted foods are more likely to develop these conditions in any case.) Some nutritionists talk of salt as if it were life threatening *per se,* and issue dire warnings that tens of thousand of deaths per year are due to its over use. Such a message may reduce health risks in developed countries, but might well be counterproductive if it is heeded in other parts of the world. There it might discourage its use and so prolong the battle to ensure no child need ever be born with cretinism.

Functional Foods

A functional food is one that offers a benefit over and above its nutritional value. Our body needs an intake of six essential dietary components in order to sustain life and regenerate cells; they are carbohydrates, proteins, fats, vitamins, minerals, and water. Many meals provide most or all of them. A functional food has to offer something extra and not just more of one of these essential components, and it must actively promote the health of those who consume it. Merely 'fortifying' a food with calcium or vitamin C does not turn it into a functional food. Most breakfast cereals have added iron, and adding more iron in the form of iron powder to make Kellogg's Special K does not make this a functional breakfast cereal.

Functional foods were first marketed in Japan where, in the 1980s, the food industry began to target an ageing and affluent population who were willing to pay premium prices for foods that might lead to a longer, healthier life. By the 1990s functional foods were becoming popular in other developed parts of the world, and in particular in the

US and Europe. By the end of the last century sales of such foods were in excess of $1 billion a year. The US even came up with an alternative name for them – neutraceuticals – which was devised by Stephen Defelice, director of the Foundation for Innovation in Medicine, based in New Jersey. Neutraceutical was coined from nutrient and pharmaceutical, but consumer research found that 60% of those questioned actively disliked the new word, while 70% liked the term functional food. Other names suggested for such products were also a bit of a mouthful: foodiceuticals, phytonutrients, designer foods, pharmafoods, and hypernutritious foods. None caught on.

Meanwhile new products were launched and advertised,[47] articles appeared in newspapers and magazines, and the subject even seemed to be gaining academic acceptability in that the University of Illinois set up the Functional Foods for Health Program with its own web site. Visit that web site today and you'll find that the program was terminated in July 2004 through lack of funding. This does not mean that functional foods were merely a passing fad, only that there was not enough funding to support study at an academic level. The idea behind functional foods is a good one and there is every reason to think that as this century progresses they will become more important.

Benecol margarine is a successful and widely advertised functional food. It was launched in Finland in 1995 by the company Raisio, and the active ingredient in this rapeseed oil margarine is the chemical sitostanol ester. Tests on rabbits in 1981 proved that this chemical could reduce cholesterol levels. Sitostanol is extracted from tall oil which is a by-product of the wood pulp industry and comes from the bark of trees. Tonnes of sitostanol are produced every year for Benecol products. Sitostanol itself is not suitable as a food additive because it does not blend well, even with oil-based foods like margarine. This will not absorb enough sitostanol to deliver the right amount to cause a meaningful reduction in blood cholesterol levels. For this reason the sitostanol is converted chemically to sitostanol ester, a form which is more soluble and so it can comprise 1 g of a normal 10 g serving of Benecol margarine.

Trials carried out in Finland showed that three servings per day of Benecol would give you enough of its active agent to lower blood cho-

47) Care has to be taken by all producers of functional foods when it comes to advertising. They are not allowed to promise cures for existing diseases, but they are allowed to say that consuming their product will have health benefits.

lesterol by 10% on average and in some of those studied it was found that Benecol reduced the 'bad' form of cholesterol, known as low-density lipoprotein (LDL), by up to 14%. The higher the level of LDL in the blood, the greater the risk of heart disease. Cholesterol is manufactured in the liver and stored in the bile duct from which it is released in order to assist digestion of the fats in our food. A lot of it is then reabsorbed from the intestines, together with any cholesterol that is part of the food we have eaten. Sitostanol is chemically very similar to cholesterol and it works by blocking the reabsorption of cholesterol.

While there is money to be made manufacturing functional foods, there are some foods that are inherently functional, none more so than fruit and vegetables, some of which contain natural chemicals that on the face of it appear to have no nutritional value, but they may have health-bringing properties nevertheless. For example, vegetables like cabbages, turnips and Brussels sprouts produce the natural chemicals cyanohydroxybutene and sulforaphane which some believe might ward off heart disease. The four main categories of functional agents of plant origin are carotenoids, flavonoids, isoflavones, and phytosterols like sitostanol. All are said to have antioxidant, anti-cholesterol, and anti-cancer properties, and prevent heart disease.

Carotenoids are fat-soluble chemical precursors to vitamin A and they protect against degenerative diseases by neutralising free radicals. Carotenoids are the bright yellow, red and orange pigments that give tomatoes, carrots, and oranges their colour. Tomatoes contain the carotenoid lycopene which is thought to prevent cancers of the breast, gut, cervix, bladder, skin, and prostate.

Flavonoids are polyphenols, and as such have antioxidant properties as well as improving the circulation and lowering blood cholesterol levels. They are found in fruits, vegetables, wines, beers, and teas. This last drink contains a range of polyphenol chemicals, some of which have anti-cancer properties, while red wine has the polyphenol resveratrol, which reduces the risk of cardiovascular disease – at least in middle class, middle-aged men. Flavonoids are also said to offer protection against cancer and allergies, although the latter claim is doubtful.

Isoflavones are phenolic chemicals found almost exclusively in soy beans and they resemble oestrogen and as such are thought to offer protection against breast, bowel, and prostrate cancers as well as counteracting the adverse effects of the menopause. (Soy bean flour has lots of other chemicals that are unique to this type of flour.) In the USA, the Food and Drugs Administration permits products with at least 6.25 g of soy protein per serving to put the following message on the packaging: "25 g of soy protein a day, as part of a diet low in saturated fat and cholesterol, may reduce the risk of heart disease."

There is a lot to be gained by eating a wide selection of fruit and vegetables each day, if only to ensure that you get the complete range of vitamins and minerals your body needs. Some vegetables have unique ingredients, for example, carrots contain falcarinol, a natural pesticide exclusive to carrots and which protects the vegetable against fungal diseases. Falcarinol is said to protect against colon cancer – and that claim is backed up by studies on rats. Other highly recommended foods include cranberries, which contains benzoic acid and relieve urinary infections, while garlic and onions have high levels of alliin and allicin that are also thought to prevent cancer and protect the heart by lowering high blood pressure and reducing cholesterol levels. Pomegranate juice is said to reduce fatty deposits in the arteries.

Animal foods can also have a functional benefit. Full cream milk and butter contain CLA, short for conjugated linoleic acid, which protects against breast cancer, and lamb also has above average levels of this type of fat. Fish oils are rich in omega-3 polyunsaturated fatty acids that some believe protect against heart disease, although this claim has been thrown into doubt by a study carried out by a team led by Lee Hooper of the University of East Anglia in England and published in the *British Medical Journal* in March 2006. This showed that those who regularly took omega-3 food supplements were no less at risk of dying of heart disease than those who did not. And it did not matter whether the omega-3 came from fish oils or linseed oil, both were equally ineffective. Despite adverse findings like this, the functional food parade rolls merrily on its way, because there are thousands of natural chemicals in all kinds of edible plants and no doubt some will be found to have health-giving properties, thereby raising them to the status of functional foods. Whether any of this will lead to healthier lives we can but hope, yet there is another route to health and that is to boost the good microbes that live within our gut.

Crap Food? Feeding the Fifty Billion

Two types of functional food are not designed to provide *us* with dietary components, but to provide food for the microbes that inhabit our gut. These are the so-called *probiotic* and *prebiotic* foods. The former are meant to introduce better bacteria into our intestines, while the latter feed the bacteria that are already there in the hope of boosting more of the good bacteria and thereby helping them crowd out the bad bacteria. Various health claims have been made for these foods.

The human gut contains around a hundred trillion living microorganisms. Indeed there are more inhabitants of our digestive tract than there are cells of our own body, of which there are around ten trillion. The weight of bacteria in our gut is around 1 kg but they are essential because they help digest our food as well as stimulate the immune system. Their numbers increase the further down the gut we go. There are as many as a thousand bacteria per millilitre (ml) of fluid in the stomach, up to ten million per ml in the intestines, and between ten billion and ten trillion per ml in the faeces in the colon. There are around 400 species of bacteria, some of which are good, such as *Bifidobacter* and *Lactobacillus*, and some are bad, such as *Enterobacteriaceae* and *Clostridium* species. Bacteria begin to colonise our gut from the moment we open our mouth as we emerge from our mother's womb.

While most bacteria are benign, some may be not so good for us – a theory put forward by Nobel laureate Elie Metchnikoff (1845–1916). His book *The Prolongation of Life*, published in 1907, suggested that the reason Bulgarian peasants were so long-lived was that they ate a lot of yoghurt containing the *Lactobacillus* species of bacteria. While his theory was popular for some years, it eventually fell out of favour in the West when the longevity of Bulgarian peasants turned out to be little more than an urban myth. Over in Japan, Metchnikoff's ideas fell on more fertile ground and in the 1930s they were developed by a medical microbiologist, Minoru Shirota, of the University of Kyoto. He searched for good bacteria that would survive passage through the harsh conditions of the human stomach with its high acid content and digestive enzymes, then survive the digestive effects of bile, finally to reach the large intestine where they could be most effective. The idea was to recolonise the intestines with healthier strains of bacteria. He found them by culturing samples taken from human faeces, and discovered a new bacteria, *Lactobacillus casei Shirota* (named after him), with which he developed a fermented milk drink known as Yakult, and which is now sold around the world. Each 50 ml bottle of Yakult contains more than 6 million of these bacteria.

The contents of our intestines have never been a popular subject for polite discussion but this part of our anatomy is crucial for healthy living and worthy of research. In the 1990s Tomotari Mitsuoka of Azabu University, Japan, was able to demonstrate how the microbes in our gut change with age and why it is necessary to keep replenishing the stock of good bacteria. The benefits of re-colonization (no pun intended!) might be to overcome various afflictions such as lactose intolerance (milk contains a lot of this carbohydrate), inflammatory bowel disease, and digestive tract ulcers. Other conditions that might benefit were said to be the diarrhoea that often follows a course of antibiotics, and even cancers of the intestines and colon.

Another popular probiotic drink is Danone's Actimel which has *Lactobacillus casei immunitas* as its active bacteria. An English company MD Foods of Leeds, produces a similar product, Gaio yoghurt, containing a bacterial culture known as *Causido*, named after the inhabitants of the Caucasus Mountains in the Ukraine from whose faeces it was extracted and who were also reputed to live long and healthy lives. Tests on Gaio yoghurt appeared to show that it even lowered blood cholesterol although these results have been challenged.

Drinks like Yakult and Actimel are called probiotic drinks, a name invented in 1989 by Roy Fuller, who describes himself as an intestinal microecology consultant and who is based in Reading in the UK. He has collaborated with Jos Huis in't Veld and Robert Havenaar of the TNO and together they preach the health benefits of such probiotics. In addition to *Lactobacillus casei Shirota* they have identified several other 'good' bacteria such as the lactobacilli *Lactobacillus acidophilus, Lactobacillus delbrueckii bulgaricus, Lactobacillus GG, Lactobacillus johnsonii,* and the bifidobacteria *Bifidobacteria animalis, Bifidobacteria bifidum, Bifidobacteria brevis, Bifidobacteria infantis,* and *Bifidobacteria lactis.* All have been recovered from human faeces and all can withstand the rigors of processing and passage through the stomach and small intestines.

With regular consumption of probiotics the composition of the gut microbes can be changed although they have to be consumed regularly if this change is to persist. The new bacteria might also make the contents of the intestines more acidic and thereby prevent disease pathogens like *Salmonella, Listeria,* and *Escherichia* from multiplying. They may also inhibit the *Helicobacter pylori* which live in the harsh environment of the stomach and cause chronic indigestion and even ulcers. Even when probiotics don't prevent illnesses they may aid recovery from diseases of the gut, such as infant rotavirus diarrhoea, a disease which kills around half a million babies a year worldwide. This was shown to be the case by a group led by Erika Isolauri at Tampere University Hospital in Finland where they fed *Lactobacillus GG* to infants who had suffered from such diarrhoea.

If you don't like the idea of consuming live bacteria then the alternative is to feed the good bacteria that you already have so that they will increase in number and crowd out the bad bacteria. Prebiotics was a term invented in 1995 by Professor Glenn Gibson of the School of Food Science at the University of Reading, England, to describe foods that will do just that. A more scientific name for them is →oligosaccharides. These are non-digestible carbohydrates which can pass through the stomach and the small intestine without being broken down, so that they reach the colon intact, there to feed the good bacteria. There are three kinds of prebiotic carbohydrates: fructo-oligosaccharides (FOS), galacto-oligosaccharides (GOS), and lactulose. FOS consists of a chain of fructose units, as many as 60 in some cases, with a glucose unit at one end, GOS consists of two linked galactose units plus a glucose unit, and lactulose consists of just a galactose and fructose joined together.

These carbohydrates may be added to all kinds of foods, such as cereals, cakes, biscuits, and health drinks. They can be extracted from things like chicory root or produced from sugar by the action of specific enzymes. A little FOS is also to be found in bananas, leeks, and wheat, and the other prebiotics also occur naturally, but no fruit or vegetable by itself can supply the 5 g of oligosaccharides needed daily to boost the good bacteria. Indeed the normal person's diet contains only about 2 g of these carbohydrates.

Mother's milk also contains several prebiotic carbohydrates that the baby cannot digest and these too feed the friendly bifidobacteria in the baby's gut. The more bifidobacteria there are in the gut the less there will be of those which cause gastrointestinal infections, namely the various species of *Campylobacter, Salmonella* and *Clostridium.* In formula fed babies these bad bacteria account for much of the microflora that are present whereas in breast-fed babies they are much less likely to be present, and this may be due entirely to the oligosaccharide levels in mother's milk. This has between 3 and 15 g per litre of these carbohydrates. Good bacteria ferment them to produce short-chain organic acids, such as acetic and butyric acids which not only lower the pH of the gut contents and act to kill other bacteria, but are also an energy source for the cells of the gut wall, enabling these to produce a thicker coating of the mucus that coats the gut and protects it. About two thirds of the body's immune system is to be found in this lining of mucus.

Infant formula feeds now contain the oligosaccharides GOS and FOS with more of the former than the latter, the ratio being around 90% to 10%. Feeding a baby the new kind of formula feed leads to levels of acid that are more in accord with what is found with babies fed on breast milk and this reflects the improved composition of bacteria in the baby's intestines.

Mother's milk also contains other protective agents, such as lactoferrin and polyunsaturated fatty acids. Lactoferrin is a protein that binds strongly to iron atoms, thereby denying them to the microbes like bacteria, fungi, and viruses which also need this metal. Polyunsaturated fatty acids are required to build cell membranes and both docosahexaenoic acid (DHA) and arachidonic acid (AA) are the omega-3 and omega-6 fatty acids known to be abundant in breast milk. DHA is believed to have a role in the development of the eye and brain as well as promoting a healthy heart. We all need these essential fatty acids which are abundant in oily fish but they can be manufactured in other ways. The main producer of AA is the Dutch chemical company DSM and they make it by fermentation using the fungus *Mortierella alpine*. Similarly DHA is made by fermenting algae, which is where fish get their omega-3 and omega-6 oils from. Some egg producers add DHA to the feed for their hens to boost the amount of this chemical in their eggs. It was already known that feeding hens flax seed and fish oil increased the amount of omega-3, and omega-6 in their eggs, which are known as Columbus eggs in the UK, and they can then be sold as functional food eggs at a premium price. In the 1990s in Japan, the company Yamazaki Baking started baking bread fortified with added DHA. There is no reason why one day DHA and AA might not be used to fortify ordinary milk so that we can all reap their benefits throughout life.

Eat Up Your Crusts!

A French stick is mainly crust, and in France it is common practice to ignore the softer inner part, and maybe that's how all bread should be eaten because the healthiest part of bread is the crust, at least according to Thomas Hofmann of the University of Munster. He reported in 2002 that crusts contains an antioxidant, pronyl-lysine. This forms from starch and the amino acid lysine when bread is baked, and it forms near the surface of the loaf, not in the soft interior. For pronyl-lysine to form the temperature has greatly to exceed 100 °C, which it does on the outside of the bread which is generally cooked at about 250 °C, but not in the middle where the water content means it cannot exceed 100 °C. Preliminary tests showed pronyl-lysine boosted by 40% the levels of enzymes that are though to prevent cancer in human intestinal cells. To what extent this would protect against cancers of the gut remains to be seen. Parents in the UK traditionally try to persuade their children to eat up their crusts by saying it will make their hair curly. The advice is sound, even it the reason is somewhat less than scientific.

Hot, Hot, Hot

According to verse 4:14 of the *Song of Solomon* in the *Bible* the chief spices are

pikenard and saffron, calamus and cinnamon,
with every kind of incense tree,
myrrh and aloes, with all the best spices.

Saffron is the dried stigma of the autumn crocus (*Crocus sativus*), once employed as a natural orange dye for cloth and used as a herbal medicine. It takes 70,000 flowers to make a pound of saffron, which explains why it is so expensive and highly valued. Today, Spain supplies most of Europe with saffron, although at one time it was grown in East Anglia in the UK when there was a flourishing textile industry there, and it gave its name to the town of Saffron Walden. Adulteration of saffron with cheaper materials, such as from the common crocus, has always been a problem despite terrible punishments for offenders, such as befell Jobst Findeker of Nuremberg who was burnt at the stake in 1444 for just this crime.

Saffron is used to colour and flavour food, the best known dishes being paella and the French fish soup bouillabaisse. There is also saffron bread, popular in the Balkans and Scandinavia, and saffron cake which is baked in Cornwall, England. Saffron has two natural chemicals that cause it to be highly valued: crocin and safranal. The former is the yellow dye and the latter gives it its savour.

What the *Song of Solomon* didn't mention was the most potent spice of all – chilli – probably because at the time it was written the only people growing chilli peppers were the native peoples of South America and even the great Solomon's influence did not extend that far. Today the kingdom of the chilli embraces the world and its grip on some countries is powerful indeed, none more so than in India. Talk of Indian food and you talk of curry. Talk of curry and you are talking degrees of hotness, not of temperature but of sensation; even a cold curry sauce can be painfully hot on the tongue. Curry powder consists mainly of turmeric blended with paprika, ginger, coriander, cardamom, clove, allspice, cinnamon, sugar, salt – and chilli powder. Of these both paprika and ginger can produce only a mild sensation of heat, whereas chilli produces a *fiery* sensation, due to the natural chemical that it contains: capsaicin.

Chilli peppers are the fruit of the *Capsicum frutescens,* a plant that is native to Bolivia, where it was cultivated for more than 7,000 years before being transported to Europe – where at first it was not well received – and thence to India where it was much better appreciated. Chilli peppers are not just a spice but they are also a good source of vitamins A, C, and E. They are rich in folic acid and potassium, and low in food calories and sodium. Nevertheless, these benefits are minor compared to the major reason for adding chilli powder to our dishes, the aim being to make the food *spicy hot.*

Chillis are hot, but how do we assess this kind of hotness? In 1912 an American chemist, Wilbur Scoville (1865–1942), devised a scale of chilli hotness by grinding up chillis, adding them to a sugar solution, and then asking a panel of testers to taste them. He then diluted the solution, and continued to dilute it, until the taste was no longer detectable. Obviously the greater the dilution required to remove its taste completely, the stronger was the original chilli.

At the mild end are bell peppers which score around 1, then come peppers like New Mexican peppers which score up to 1000, while Jalapeno peppers score around 5000, Tabasco and Cayenne peppers are much hotter at 30,000 to 50,000, Scotch Bonnet and Thai peppers go over 100,000 and finally there is the truly impossible Red Savina Habañeros which clocks up a massive 577,000. There are even claims from the north-eastern state of Assam in India that the Naga Jolakie chilli scores even higher at 855,000 units. On Scoville's scale pure capsaicin rates a massive 15,000,000 units. Capsiacin's effects are particularly marked on humans. Other mammals are also affected although there are some mice for which it does not register, and birds will quite happily eat even the hottest chilli peppers, including the dreaded habañero. Indeed birds act to disperse the seeds of the plant while at the same time its hotness deters most mammals from eating them.

The capsaicin molecule has a framework consisting of a chain of nine carbon atoms linked via a nitrogen atom to a benzene ring. (The active molecules in paprika and ginger are somewhat similar but with shorter and longer chains respectively and different chemical groups attached.) There is nothing in the molecular structure to indicate how these complex molecules will interact with receptors on the tongue or why capsaicin is particularly hot. Capsaicin is insoluble in water but soluble in alcohol and organic solvents, the latter being used to extract it from chilli peppers which can contain up to 1% capsaicin. Capsaicin is also fat-soluble because it has oil-like features.

Our tongue can immediately detect whether a food or a drink is hot or cold and it can judge the degree of heat or coldness, which it does through two kinds of receptors. The reason capsaicin registers as a hot sensation on the tongue is that it activates the same receptor VR1 (short for vanilloid receptor type 1) as does heat. Capsaicin causes the sensory nerves to send a pain signal to the brain by binding to a receptor that opens the channels through which calcium escapes and these in turn activate a pain-sensing nerve cell to release a neurotransmitter that triggers the pain signal. The capsaicin molecule binds tenaciously to the receptor and continues to hold the channel open until it is dislodged. The same thing happens when you put something with a temperature above 43 °C into your mouth, until it cools down and the channel closes.

There is a similar link between chemicals that produce a cool sensation in the mouth and food that is cold. Chemicals such as menthol and eucalyptol activate a different type of receptor called CMR1 (short for cold menthol receptor type 1) again causing calcium ions to flow out of those cells and again triggering a pain-sensitive nerve cell, but this time one that registers cold. The same thing happens when you put something cold in your mouth. We know that CMR1 is a membrane protein with channels through which metal ions can pass in and out of the cell, but it has a tap that can only be opened by a certain physical sensation (cold) or a chemical activator such as the cool tasting menthol. The heat detector receptor VR1 works in a similar way but its protein structure has yet to be deduced. Together VR1 and CMR1 receptors provide the body with a thermometer capable of sensing temperatures in the range 8 °C to 60 °C.

Over-stimulated pain receptors eventually release endorphins, which are natural pain-killing molecules. VR1 receptors can lose their responsiveness after prolonged exposure to capsaicin which is why those who eat a lot of spicy food build up a tolerance to chilli. This is also why capsaicin is used in formulations designed to ease pain, repeated application of which desensitise the nerves. Regular application of a capsaicin cream to aching joints can relieve pain and increase flexibility. Chemotherapy for cancer patients often results in oral pain and sucking capsaicin-laced butterscotch has proved effective.

Beyond its role as a painkiller, capsaicin has other benefits. It speeds up metabolism and lowers cholesterol levels, whilst it has antibacterial properties that may protect against peptic ulcers by killing the bacterium *Helicobacter pylori* which causes them. Chilli pepper's popularity in hot countries may in part be due to its ability to destroy bacteria of the kind that can cause food to go bad and this seems to be borne out by the fact that the countries with the spiciest cuisines are the hottest ones, namely Thailand, the Philippines, India and Malaysia, while those which use the least are the coldest, namely Norway, Sweden and Finland.

Pepper Spray

Capsaicin is the active agent in the pepper sprays which are used by some police forces to disable and arrest aggressive individuals. Sprayed at the eyes they produce an intolerable sensation of burning although no permanent harm, the effect wearing off after about 30 minutes. Capsaicin is also the ingredient in sprays used to protect crops of fruit, grapes and vegetables, and it has been incorporated into marine paint as a way of deterring barnacles from growing on the bottom of ships.

Several other spices have even better antibacterial properties than chilli. Garlic, clove, and cinnamon have chemicals that destroy the deadly bacteria *Escherichia coli* 0157:H7. The chemicals responsible for the activity of these spices are cinnamic aldehyde in cinnamon, eugenol in clove, and diallyl thiosulfinate in garlic. The spices with the strongest antimicrobial properties are (in order of strength) first garlic, next onion; then allspice, followed by oregano, thyme, cinnamon, tarragon, cumin, cloves, and lemon grass. The spices with least antimicrobial effect are parsley, cardamom, pepper, and ginger.

There are other spices that may be more than just flavoursome and which we might also consider to have functional aspects. Curcumin is the bright yellow chemical in turmeric, one of the main spices in curry powder, and it has been shown to have anti-cancer properties because it inhibits an enzyme APN (short for aminopeptidase N) that cancers use to ensure their own blood supply. APN is a zinc-containing enzyme that breaks down proteins at the cell surface and this enables cancer cells to invade the space of neighbouring cells. Professor Ho Jeong Kwon at Sejong University, Seoul, screened 3,000 molecules for activity against APN before he came across curcumin. It is now undergoing clinical trials for colon cancer. Another benefit is that it is a drug that can be taken orally and it doesn't appear to have any side effects.

Scientists at the University of Michigan Medical School, in collaboration with those at the Indian Institute of Science in Bangalore, have carried out tests on curcumin that show it to inhibit the drug-resistant forms of malaria and reported their findings in the *Journal of Biological Chemistry* in December 2004. Mice infected with the related parasite, *Plasmodium falciparum,* which causes rodent malaria, were fed curcumin and this reduced the number of parasites in the blood by as much as 90%, and completely protected more than a quarter of mice to whom it was given. Whether it could be a treatment for human malarial infection remains to be seen.

Food will always be a topic of interest, but most people know little of its *chemical* composition and yet that is the key to understanding what it can and cannot do for us. Food chemistry is an important branch of the subject and food chemists are employed by many companies, working to ensure that what we eat is safe and nutritious. Because what they do is not understood, they are often seen as interfering with the food and some commentators see this as highly suspicious. Occasionally there are alarms about 'chemicals' in food and sometimes the culprits are not the ones that come from the chemical industry, but the ones we make ourselves in our own kitchens, as the following story shows.

Issue: How Did the Toxic and Cancer-causing Chemical Acrylamide Get into Everyday Foods?

Acrylamide[48] has been manufactured since the 1950s and is used to make things like dyes, adhesives, and water-tight sealants. While most of these products contain no acrylamide, there are traces of this starting material in sealants. These are used to seal sewer pipes and manhole covers and this led in 1991 to a ban on such products by the US Environmental Protection Agency because acrylamide can penetrate the skin and even diffuse through rubber gloves. As such it presented a threat because it is a neurotoxin, in other words it can poison the nervous system of those exposed to it. (The ban was withdrawn in 2002 when newer and better protective clothing was developed.)

In 1997, work on a railway tunnel that was being built in southern Sweden was being delayed because water was seeping into it. To prevent this, large amounts of an acrylamide-containing sealant were applied to the tunnel walls but acrylamide began to leach into the water table and affect salmon in local rivers. It also affected several of the men working in the tunnel, who experienced symptoms such as numbness in hands and feet, headaches, and dizziness. Blood samples were taken from the men and analysed at the University of Stockholm and they showed the presence of acrylamide.

Because acrylamide is not a chemical normally found in blood it was necessary to take samples from people who had not been exposed to acrylamide to act as a control group. Much to the surprise of the analysts, all of them had measurable amounts of acrylamide in their blood. While smokers were expected to have some acrylamide in their bodies, because it is a component of cigarette smoke, non-smokers should not have had any, and yet there it was. Blood samples were collected from more people, including those from other countries, and they too had this dangerous chemical circulating around their bodies. Sweden's National Food Administration were alerted and deduced that the acrylamide was coming from the food people were eating. But how was this toxic chemical getting into food?

It turned out not to be an environmental pollutant from the chemical industry, as some suspected, but to be a natural product formed when food was prepared, and it is particularly prevalent in all kinds of fried potato. The more chips, crisps, and fries a person ate the more acrylamide was in their blood.[49] Some potato crisps were found to have in excess of 12,000 ppb although this was exceptional. The amounts of acrylamide in foods are listed on the US Food and Drugs Administration web site www.cfsan.fda.gov/~dms/acrydata.html although this refers only to foods available in America; however, analysts in other countries reported similar levels. The FDA found levels of acrylamide in crisps varying from as low as 117 to as high as 1970 ppb and French fries ranged from 21 to 1325 ppb. Other foods had much less, although some biscuits

48) Chemical formula $CH_2{=}CHCONH_2$.

49) In the UK chips are chunky fingers of potato freshly fried, fries are thin and made from mashed potato that has been squeezed into thin strips and fried, while crisps are fried slivers of potato eaten as a snack or with dips. In the USA crisps are known as chips.

had in excess of 1000 ppb, and there was even one breakfast cereal with more than 1000 ppb. Baby foods had almost none at all. Bread had as little as 20 ppb although when this was toasted the amount could rise to 200 ppb.

Why was acrylamide being formed? The answer was provided by Bronislaw Wedzicha and Donald Mottram of Reading University in the UK who proved that it came from the amino acid asparagine reacting with carbohydrate. This reaction is of the kind which causes browning when foods are cooked. The longer food is cooked, and the higher the temperature, the more acrylamide was formed. Potatoes, in which asparagine is the dominant amino acid, gave most acrylamide when fried at 185 °C.

Sweden's National Food Administration went public with their findings in 2002 and the news naturally caused alarm, especially when a leading member of that organization said: "It seems reasonable to assume that ... several hundred of the annual cancer cases in Sweden may be attributed to acrylamide." Hundreds in Sweden implied hundreds of thousands worldwide. Acrylamide was already known to cause cancer in laboratory rats although these are specially bred to be susceptible to the disease. Such susceptibility doesn't automatically mean it will cause cancer in humans, indeed it rarely follows, and an investigation published in the *British Journal of Cancer* in 2003 said that there was no link between the amount of acrylamide in the diet and cancers of the bowel, kidney, or bladder, which are the organs most likely to be affected.

And there was another problem: how much acrylamide you detect depends on the way you analyse for it. In fact there was no foolproof way of measuring it, as the pages of the leading journal *Analyst* were to prove. One method based on slow solvent extraction, and used in Sweden, gave levels of acrylamide seven times high that the more general method based on rapid extraction. One batch of potato crisps had more than 14,000 ppb according to the slow extraction method, whereas it had only 2,000 ppb when analysed by the more general method. Analytical chemists in Canada, the UK, the USA, and Japan criticised the slow method because it is itself capable of forming acrylamide.

The accepted safe intake of acrylamide for the average person is 14 micrograms per day, although the World Health Organization estimates that a Western-style diet provides 70 micrograms per day. So should we worry about this? The answer is probably no. It would appear that our body's natural detox defences are more than capable of disposing of this amount quite easily so we can continue to enjoy fried potatoes. If you are still worried, then you should always eat protein with your chips or fries because this can bind to acrylamide and prevent it entering the cells of the body.

5
Better Living (III): Minor Metals for Major Advances

A small arrow printed before a word in the main text indicates that there is more information on that topic in the Glossary.

News from the Future

Global Times News, 21 March 2025

New Village with Zero Energy Bills

The new Dolphin retirement village at Gomera, in The Canary Islands, was officially opened yesterday by the Minister for the Environment who praised the fact that it was entirely self-sufficient in energy. All cottages are equipped with fitted kitchens and entertainment facilities and are fully air-conditioned thanks to SuperGlass windows and roofs.

SuperGlass adjusts to the level of sunlight falling on it, allowing light and warmth to enter when the weather is cool but when the weather is hot it generates energy which can be used to work air-conditioning – and the new glass is also self-cleaning.

Dolphin village generates enough electricity to power all its buildings plus a desalination plant that provides its 500 dwellings with two million litres of water per week, enough to supply all that is needed for washing, cleaning, and toilets, as well as irrigating the large gardens and fruit trees, which are mainly oranges and avocados, growing on its terraced hillside overlooking the Atlantic. The village also has cafés, bars, heated pool, leisure club, shops, health centre, and golf course.

Page 7: Recycling SuperGlass will be prohibitively expensive claims environmental group.

Most of the chemical elements are metals and many of them are little used; some are so rare that only the man-made variety exists on earth, and that applies to the radioactive elements beyond uranium, which is element number 92 in the periodic table. It is also true for lighter elements such as technetium (element 43) and promethium (element 61). Yet such elements have their uses; technetium in medical diagnostics, promethium in miniature batteries for pacemakers. Hopefully, most people will never need to encounter either metal, but some radioactive metals should be a part of everyone's life and especially

Better Looking, Better Living, Better Loving. John Emsley

ISBN 978-3-527-31863-6

americium (element 95) which is in all smoke detectors. When its stream of alpha rays is blocked by smoke particles then the detector sounds an alarm.[50]

The metals in this chapter are not ones we come across as the metals themselves but as their chemical compounds. You may not even recognise their names or if you do you may not think of them as metals. The business world categorises metals into four groups: base metals, precious metals, minor metals, and rare earth metals. It is some of those in the minor metals category that concerns this chapter, and these are becoming more and more important. There is even a Minor Metals Trade Association whose members seek out sources of them and bring them to market where they find use in ways that already impinge on everyday life, such as mobile phones, solar panels, radar, smart windows – and traffic lights. The metals in question are gallium, indium, titanium, ruthenium, and cadmium.

Solar Panels Provide Photovoltaic (PV) Power

Electrons moving through a wire are the most versatile supply of energy, but generating this electricity and getting it to where it is needed consumes vast amounts of non-renewable fuel, and a third of its power has been lost to the environment by the time the electric current reaches its destination. Ideally we should generate electricity from sustainable sources of energy and as near to the point of use as possible. Solar panels meet both these requirements. Harnessing the rays of the sun in order to mobilise electrons is something plants learned to do a long time ago, using chlorophyll to capture a photon of light and kick an electron into action as part of the photosynthesis by which a leaf can make the various chemicals the plant needs. Nature does it and so can we. Given the right material, a solar panel can similarly capture a photon of light to start an electron moving and given enough of these we have an electric current for our use. Nature, however, isn't going to make it easy. For a solar cell to convert sunlight to electricity it requires a semiconductor which can absorb solar energy and use this to generate negative electrons and positive holes. These need to

50) These alpha rays do not pose a threat to humans because they are contained within the detector unit and cannot escape from it. Indeed alpha rays can be stopped by a sheet of paper.

be separated so that they move in opposite directions thereby producing an electric current, and this is what happens at a semiconductor diode junction. The result is photovoltaic (PV) power.

First generation PV cells are made of crystalline silicon and they account for most of the solar electricity currently generated, but their efficiency in turning sunlight to electricity is less than 20%. Second generation cells are just as efficient but use much less silicon or are made from alternative materials such as copper-indium-diselenide. Third generation cells use a combination of semiconductors as in the multijunction solar cells (also known as tandem cells). These have a top layer of gallium indium phosphide (GaInP), a middle layer of gallium arsenide (GaAs), and a lower layer of germanium (Ge) and they are able to harvest sunlight across more of the sun's spectrum and achieve power efficiencies of 39%. In June 2005 the company Spectrolab, a subsidiary of Boeing based in Sylmar, California, was claiming this level of efficiency for the multijunction cells it manufactures. These third generation cells can maximise their output by concentrating the rays of the sun on to them while tracking its path across the sky.

On a global scale, the electricity we generate in thousands of power stations is enormous. Even so, it pales in comparison to the amount of energy the Earth receives every day from the Sun. There is enough sunlight falling on the Earth in one hour to supply all the world's energy needs for a year. We can harvest some of this using solar panels, but these generate less than 0·25% of the 4000 →gigawatts (GW) which the world needs. A Greenpeace report in 2005 spoke optimistically of raising this to 20% by 2040. What might prevent this, and even attaining a more realistic target of 5%, is the supply of elements that would make it possible, and these include the little known metals gallium and indium. (The likely shortage of indium is discussed at the end of this chapter.) Our reluctance to invest in this 'free' energy is paradoxically due to its high cost: the energy generated by solar panels is several times more expensive than energy generated in other ways. Despite this, the world is investing in solar energy and in 2005 the generation of electricity by solar cells amounted to 5 GW worldwide. The current target is to reach at least 15 GW by 2010, by which time the global demand for electricity is calculated to be 4500 GW. (The US Government's Energy Information Administration predicts this will rise to 5000 GW by 2015, and 6000 GW by 2025.)

Most solar panels are to be found in Germany, Japan, and the US which together account for four fifths of the world's solar power generation. Indeed Germany installed more photovoltaic power in 2005 than the rest of the world put together, although most solar cell production is based in Japan. That country aims to meet 50% of its electricity supply from PVs by 2030 when there should be an array of solar panels on most roofs. They might well achieve this because they have some of the largest solar cell manufacturers in the world: Sharp leads the field and in 2005 it produced enough to generate 0.43 GW, Kyocera produced 0.14 GW, Sanyo 0.13 GW, and Mitsubishi 0.10 GW. Other major manufacturers are the German company Q-Cells (output 0.16 GW), Schott Solar (0.10 GW), and BP Solar (0.09 GW). Spain, China, Greece, and Italy are also investing heavily in solar energy, with Spain leading the group by installing 0.10 GW per year. The UK by comparison generates only around 0.006 GW of its electricity this way, but then that country is situated on a northern part of the globe and the sun's rays are often hidden by clouds. As we shall see, these disadvantages can be overcome, and solar power generation is on the move even there, where two factories now manufacture solar modules.

Despite their cost, solar panels are essential for some communities. For example, in September 2001, more than 150 villages in the remoter parts of the Philippines were supplied with solar generators, and the villagers now enjoy the benefits of electricity for homes, schools, clinics, water treatment plants – and Christmas tree lights. The scheme came as a result of collaboration between the Spanish and Philippine governments and cost $50 million. California is also a heavy investor in solar energy. It has some very innovative structures such as that at the California State Fair where 1000 parking lot canopies are covered with solar panels which produce 0.5 megawatts (MW) from sunlight that would otherwise be roasting cars.[51] An equally large array covers the roof of the Toyota building at Torrance, and this too generates 0.5 MW. In January 2006, the California Public Utilities Commission started an initiative to boost solar energy in the state with the target of installing 500 MW per year by 2015, and encouraged by $3.2 billion of incentives.

51) There is a certain irony in generating solar energy by providing shade for parked cars, which are probably the world's biggest wasters of energy.

Like all forms of power generation, solar energy systems are prone to being somewhat less than totally reliable, as events in Spain have shown. In 1994 a solar array was set up at the village of La Puebla de Montalbán in the central region and it was designed to feed the national grid with 1 MW of power. In fact it delivered 0.85 MW. This was an experimental solar station with three types of panels, two fixed and one designed to track the path of the sun across the sky. Most PV panels are of the flat bed type and these cover a large area and are static, although angled to catch the maximum amount of sunlight. Following the sun means a loss in harvested energy because a proportion of this is used to move the heavy array. During the 10 year trial there were losses of power due to faulty wiring, dust on the cell windows, burned out fuses, the disintegration of some panels, not to mention the slight problem of a thief who stole a few. However, of the 8000 solar cells only 50 of them broke down completely, less than 1%, which over a ten-year period is reassuringly few. Some loss of power is inevitable in any case, due to the fact that the silicon itself decays by about 1% a year. (Other photovoltaic materials do not suffer from this defect.)

The photovoltaic effect was first demonstrated by the young French physicist Alexandre-Edmond Becquerel (1820–1891) as long ago as 1839. It remained little more than a scientific curiosity for the next hundred years or so, and there were two reasons for this long gap between discovery and application: high cost and low efficiency. The first solar cells appeared in the 1950s and they were fitted to satellites that were put into orbit round the Earth. They were made of pure polycrystalline silicon, which is extremely expensive to produce, and they could only convert 4% of the sun's rays to electricity. Silicon still accounts for around 90% of the solar cells in use around the world, but they now have conversion efficiencies of about 15%. Silicon can absorb →light rays between the violet and the near infrared, but it does not do so at all wavelengths, and where it does, it does so inefficiently. Its output is sufficient for many uses, but if solar cells are to make a viable contribution to world energy needs then their efficiency needs to be much better than 15%, or they must be made from cheaper materials. The global shortage of silicon wafers is currently a brake on PV cell production, despite attempts to increase the supply and to use that which is produced more sparingly by cutting it into thinner films. Japan produces 95% of suitable grade silicon and the reason the industry flourishes there is that the Japanese Government decided to invest in solar energy following the Oil Crisis of the early 1970s.

Photovoltaic cells use semiconductor silicon in two layers, an electron-rich layer on top of an electron-poor layer. These produce an electric field at the junction between them. The electron-rich layer is silicon doped with a few atoms of phosphorus to give it a surplus of electrons and is known as the n-layer (negative layer), while the electron-deficient layer is silicon doped with a few atoms of boron to give it a deficit of electrons and this is known as the p-layer (positive layer). The n-layer releases electrons when excited by photons of light; the p-layer in effect moves 'holes', which are the gaps in chemical bonding where electrons are missing. Together these create a voltage difference that causes current to flow. Solar cells need to be connected in series to generate power for practical power applications, and the current they produce is direct current but this can easily be turned into alternating current.[52)]

Silicon accounts for 60% of the cost of a solar cell.[53)] Given that silicon is the second most abundant element in the Earth's crust – oxygen is the first – why is it so expensive? Semiconductor grade silicon has first to be made ultrapure which is why it has to be manufactured under 'clinical' conditions free of all possible contaminants. In 2004 and 2005 there was only a limited amount of polycrystalline silicon available for the solar cell market which resulted in many fewer solar panels being installed in those years. The price of pure silicon rose from $9 per kg in 2000 to $200 per kg by 2006. (Things should change when new and large manufacturing facilities come on-stream in China later this decade.) Tiny amounts of equally purified phosphorus or boron have then to be incorporated into the silicon before the whole is cast into large ingots weighing 240 kg. These ingots are cut into 25 brick-shaped blocks 125 mm by 125 mm and each is sliced into 400 wafers (250 microns thick) using a fine cutting wire so that each ingot eventually produces 10,000 of them. The hoped-for generation of 5 GW of solar energy by 2010 will require production of 30,000 tonnes of solar-grade silicon per year, ideally at a cost of less than €35 per kg. By slicing the crystalline silicon thinner, to between 100–130 microns per wafer, it will, in theory, be possible to manufac-

52) Direct current is more efficient but society is geared up to using alternating current at the present time.

53) Other materials account for around 20%, interest on borrowed capital for around 15%, and labour for only 5%.

ture enough silicon to hit this target. The industry is also hoping to reach an energy conversion efficiency of around 17% for mass produced silicon-based cells by 2010.[54)]

Research is also focussed on alternative materials for photocells based either on elements which are more efficient than silicon but dearer, or on organic semiconductors and carbon nanostructures, which are much cheaper but less efficient than even the poorest quality silicon. Each approach has its attractions and its drawbacks.

Other metal-based semiconductors are potentially much better PV materials than silicon because they can more closely match the energy of the incoming light rays; these are gallium arsenide (GaAs), indium phosphide (InP), and cadmium telluride (CdTe). Of these only the last has so far made significant inroads into solar power generation and that generated only 0.013 GW (13 MW) in 2004, but then again it was only in 2002 that solar panels made from CdTe exceeded the magical 10% figure for energy conversion. There is unlikely to be any economic restraint to using cadmium because this metal is an unwanted by-product of zinc refining, but it may be opposed on environmental grounds because this metal is dangerous and many of its former uses, such as in galvanising iron, have been banned. A large amount of cadmium comes from zinc mining but only a small fraction of this would be needed for solar panels and as such it would present no threat to human health.

The material which might overtake CdTe in terms of widespread application is copper indium diselenide (CIS)[55)] which is being heavily backed by the oil giant Shell. Shell's efforts are mainly directed at producing CIS thin film solar panels, and their ST40 product can generate a peak power of 40 watts at 16 volts and is designed for rural and industrial applications. The company is so sure that it has a winner that its solar pack comes with a 10 year warranty. It is claimed to work under low light conditions, to have a high temperature tolerance, and its glass surface can withstand all kinds of weather including hail storms. A cell made of CIS at the US National Renewable Energy Laboratory in Golden, Colorado, has achieved an energy efficiency of al-

54) There are two kinds of efficiency of photovoltaic materials: quantum efficiency and power efficiency. The former is the efficiency with which a material converts photons of light to negative electrons and positive holes, the latter is the real efficiency that can be extracted as electric power, and that is far less.

55) Chemical composition $CuInSe_2$.

most 20%, and well above the 6% reported for this PV material back in the mid-1970s when it was first made at the University of Maine. Replacing about a quarter of the indium in CIS with gallium to give copper indium gallium diselenide (CIGS) not only reduces the amount of the more expensive metal, but also produces an alloy that is easier to manufacture, and appears even to improve the efficiency.

Another cadmium semiconductor is cadmium sulfide (CdS) and this is being manufactured on a small scale and incorporated into panels that are a metre square and deliver 91 watts, at an efficiency of 11%. The panels are made from commercially available glass 3.2 mm thick with a coating of indium tin oxide on to which is applied a layer of CdS followed by a layer of the CdTe (p-type).

Thin film photovoltaics promise to be a major contributor to electricity production. The silicon for this is amorphous silicon which is radically different from crystalline silicon in that it does not need to be cast into ingots and it can be applied to a substrate such as glass or, better still, a flexible material so that the completed panels can be moulded to any surface. Thin film photovoltaics are now rolling off the machines at various factories around the world and some come with adhesive backing so they can be bonded to roofs such as that at the Stillwell Avenue subway station in Brooklyn, New York. Research into thin film PV is being mainly funded by defence contracts because the military sees the potential of replacing the heavy storage batteries which many armed personnel have currently to carry in order to power sophisticated equipment. Already there are rucksacks with thin film solar panels that backpackers can use and which will recharge their mobile phones and other small devices.

The alternatives to metal-based semiconductors are organic semiconductors, but they struggle to reach efficiencies of 5%, although there are hints in the scientific literature of polymers that are comparable to silicon in their semiconducting properties. Even so, polymers rely on double bonds along the backbone of the polymer to provide the structure along which electrons and 'holes' can move, but double bonds are always susceptible to oxidation by the oxygen of the atmosphere so they have to be protected.

The solar energy world finds itself on the horns of a dilemma, having to choose between metal-based solar cells, which perform well but are costly, and organic solar cells, which are cheap but inefficient. One answer is to combine the two types of semiconductor to produce a hybrid material which has the benefits of both cheapness and high efficiency, and this has been shown to be feasible by Paul Alivisatos of the Department of Chemistry at the University of California, Berkeley. His solar cell materials are composed of 20% cadmium selenide (CdSe) in the form of nanosized particles and 80% of fibres of the conducting polymer P3HT,[56] and these cells have a power efficiency of around 2%. Progress has been made in raising efficiency by stacking the polymer chains more tightly, thereby improving electron flow and this has been achieved in a cell composed of P3HT as the electron-rich layer with the electron-poor layer being a derivative of the cage molecule C_{60}. This curious molecule consists of a cage of 60 carbon atoms and it too has PV potential. In the past, organic photovoltaic cells have been held back by the low mobility in transporting electrons and holes. Carbon nanotubes are able to improve this and composites of polymer and carbon nanotubes are showing promise.

When the sun shines all is well for solar cells, but what happens when the sky is overcast? In some northern countries the sun may not shine for days on end, and in winter the hours of daylight get shorter in any case, to as little as 8 hours in many large cities in northern Europe. When there is no direct sunlight then the answer is the Grätzel cell which can operate using only daylight. Michael Grätzel at the Swiss Federal Technology Institute of Lausanne made the first such cell in 1991 and it relies on a compound of the rare element ruthenium capturing a photon of light and using this to excite electrons which are then passed on to a layer of titanium dioxide crystals and thence to an electrode. The Grätzel cell is as efficient as a solar cell because it can absorb light coming from any direction, and it is particularly good at absorbing the blue end of the spectrum which is the light that best penetrates clouds.

Grätzel cells are also known as dye-sensitised solar cells (DSSCs) and they rely on a liquid electrolyte to transport the charge, but the solvents for this have to be organic liquids which are not durable or give low conductivity. Ionic liquid crystals work better and these can self-assemble to form conductive pathways leading to better conduction, as discovered by Shozo Yanagida at Osaka University, Japan. Grätzel cells are much cheaper than conventional silicon-based photovoltaic cells but their conversion efficiency is still less than 10%. Grätzel has

56) P_3HT is short for poly(3-hexylthiophene).

developed a new dye based on a ruthenium compound, which he refers to as K-19. Unlike earlier dyes, which lose sensitivity or are not stable when heated, the new dye can survive at 80 °C for 1000 hours and even under these extreme conditions it lost only 8% of its performance. If DSSCs are to become viable then they must be able to operate for 20 years in outdoor conditions, which mean sustaining 100 million operations of each dye molecule. The Swiss firm Solaronix, based near Lake Geneva, is developing DSSCs based on Grätzel's work and says cells can deliver electricity at a third less cost than that of silicon-based solar cells.

Thirty years ago, in the 1970s, concentrated photovoltaic power (CPV) was demonstrated and it can have an efficiency of more than 30%. What it offers is cheaper electricity and whereas a normal flat-plate PV cell costs $9 per watt fully installed,[57] CPV systems can cost half as much. CPV systems use lenses to focus sunlight onto a much smaller amount of photovoltaic material, and the lenses used are Fresnel lenses which are flat with miniature circular grooves which focus light on to a central area. In a well-designed cell the light falling on an area of 100 cm^2 can be focussed on to 1 cm^2 of PV material. One slight disadvantage is that a lens transmits less than 100% of the light which falls on it, although most can transmit at least 85%. A bigger disadvantage is that lenses cannot focus diffuse sunlight and even when the sun is shining the solar collector has to track exactly its progress across the sky in order to keep the sunlight focussed directly on to the cell and to maximise the efficiency it must not deviate by more than a degree. There is yet a further drawback to CPVs in that they also concentrate the heat of the sun and as the temperature rises the solar cell's efficiency falls, so the cell has to be mounted on something like copper to conduct the heat away.

The main attraction of CPV is that it uses much less of the solar cell material which means that these can be made of expensive semiconductors like GaAs. Concentrix Solar GmbH of Freiburg, Germany, is a spin-off from the Institute of Solar Energy there and its all-glass Fresnel lens concentrator focuses the equivalent of an area corresponding to 500 suns on to tiny 2 mm^2 solar cells mounted on copper heat sinks – and efficiencies in excess of 25% are achieved. A similar '500 suns' light concentrator has been developed by Sharp and Diado Steel of

57) The panels themselves cost around $6 per watt.

Japan and that focuses on to GaAs cells which are 7 × 7 mm in size and achieves a conversion of more than 35%. Whitfield Solar is a company set up by George Whitfield of the University of Reading, England, and his silicon-based, intelligent-tracking, CPVs are designed specifically with low cost in mind, which means a minimum use of materials and maximum ease of manufacture.

Much is promised from solar energy and there are already scores of such mega arrays around the world with more to come. Portugal is building the world's largest solar PV installation at Moura in the south-eastern Alentejo region. This will be known as the Girassol ('sunflower') project and is planned to come on stream in 2009 when its 350,000 solar panels, covering an area of 112 hectares, will generate 62 MW of power capable of serving the needs of 20,000 homes. A factory to manufacture the solar panels is being built by BP Solar and this will provide jobs for 240 locals. The overall cost will be around €250 million. To date the largest solar scheme has been the Bavarian Solarpark in Germany which has a generating capacity of 10 MW and covers 26 hectares. It has 57,600 panels that track the sun and is located at Muehlhausen, Guenching, and Minihof. It was built on farmland that is no longer in use and it cost €50 million. Its solar panels were made by PowerLight, a company set up by Thomas Dinwoodie in 1991, and which is now part of Microsoft. This is also a leading player in producing solar systems that are being installed on roof tops around California and which can feed into the local grid. The solar tiles are mounted on polystyrene which acts as an insulator thereby making the underlying building more energy efficient as well.

A concerted effort might well bring great benefits as the following news item from the future shows:

The Ghana Sun, June 2025

Sunshine City Switched On!

At a special ceremony attended by dignitaries from all over Africa, the President of Ghana threw the switch that connected solar panels on the roof of the capital's parliament buildings to the Accra Solar Energy Board, thereby completing the scheme to generate all the city's electricity from solar sources.

"Accra is now the first city on Earth with more than a million inhabitants that is totally powered by sunlight" he said. "The city has one of the best electricity storage systems in the world and this can store up to a week's supply of power. Although many thought it would be impossible to meet all of Accra's electricity demands from solar energy this has been achieved thanks to a concerted effort by residents in all localities who fitted solar panels to their roofs and outbuildings. I thank them all."

Public buildings, shops, factories, warehouses, and office blocks have undergone a massive window replacement programme, and even north-facing surfaces have been fitted with Grätzel panels which can generate electricity even when not in direct sunlight. The cost of the ten-year programme was funded from a levy on cocoa and palm oil exports which are the main source of the country's wealth. Home owners were able to claim 75% of the cost of installing solar panels from the levy.

The widespread deployment of solar cells is hampered by a few problems. Their cost is still too high and will have to come down to around \$2 per W before people would invest in them, and their efficiency is too low, which means that a large area has to be covered to generate a worthwhile current. Whether that cost really is too dear will depend on alternative energies and these are rising in price as demand around the world increases. Currently solar cells also need the rare element indium and this may become a limiting factor and we will consider the issue at the end of this chapter.

Solar energy can be used in other ways, and is already used on a large scale to heat water for domestic purposes. More than 40 million homes around the world are equipped with such water heaters on the roof. This too would save enormous amounts of energy in countries like Ghana, and is saving energy in many hot countries like Cyprus, Greece, and Turkey, where almost every home has a hot water unit on its roof. In fact the solar thermal industry is much bigger than that of the photovoltaics industry although many imagine it to be the poor cousin. In 2004 solar thermal capacity increased by 9 GW in that year alone and most of that was installed in China where there is a staggering demand for solar water heaters. They now account for 12% of that country's hot water. What is also noteworthy is that solar thermal units in China cost far less than those in Europe.

Generating electricity via water heating is also possible provided you concentrate the sun's rays so as to create a pressure of steam and use that to drive a turbine or a Stirling engine. Efficiencies are of the order of 20%. The Stirling engine was invented by 26-year-old Robert Stirling (1790–1878) in 1816. It works by heating up the fluid in the hot cylinder, thereby causing it to expand and drive a piston, then passing it to a cold cylinder before being compressed by the piston and forced back into the hot cylinder. Stirling engines are silent and efficient but costly to produce. Solar thermal power requires the sun's rays to be focussed sufficiently to create temperatures in excess of 500 °C and there are some in operation around the world. Dish-shaped parabolic mirrors can be made of rigid lightweight stainless steel only 0.28 mm thick, and an array of such collectors can easily generate 10 kW of power and at a cost much less than that from solar panels. The dishes need to track the sun as it moves across the sky.

Solar panels embedded in windows, of the type reported in the 'News from the Future' at the start of this chapter, should be a feature of many buildings by the middle of this century. Then of course they will have to be kept clean even when they are not easily accessible – which brings me to the next topic in this better living chapter and one which will remove the need for this chore.

Glass is Green

This section is about an age-old material, glass, which will undoubtedly be the building material of the future – and for several reasons. The transparency of glass makes it the ideal medium for solar cells, and embedding these in a window would seem a logical step to take and such material is now being manufactured.

PowerGlaz is the name of the new architectural material designed for use as facades and roofs of buildings. Its makers say cladding a surface with PowerGlaz is no dearer than cladding it with polished granite or marble, so it is rather expensive. The panels contain an array of 12.5 cm (6 inch) square silicon-based solar cells laminated between two layers of glass 3.3 × 2.2 metres (11 × 7 ft) in size, and they can deliver around 900 watts of power when exposed to the midday sun. The panels are manufactured at the Romag factory at Consett in County Durham, England, which has the capacity to make 7,000 such panels a year, producing a total of 6 MW of photovoltaic power. If the roof of a normal house were made of such panels then it would generate enough electricity to be self-sufficient. The International Business Centre at nearby Gateshead is one of a number of public buildings now fitted with PowerGlaz panels, in this case with 36 of them, and they generate 34 kW of electricity, enough for that building's needs on a sunny day. PowerGlaz's future seems assured.

Glass has been around since the time of the pharaohs although it was the Romans who first used it for windows. Very little Roman window glass has been found on archaeological sites, however, and the reason is that it was too expensive a commodity to throw away when broken; it was saved so that it could be re-melted and used again. The ingredients for glass were abundant. The basic raw materials were sand (silicon dioxide) and sodium carbonate, which occurred naturally in Egypt, plus lime (calcium oxide) made by roasting limestone in lime kilns, and potassium carbonate which is the main constituent of wood ash. Heat a mixture of these to a high enough temperature and they melt and react chemically to form a metal silicate which, when it cools to around 600 °C becomes a transparent solid, known scientifically as a super-cooled liquid. Glass resembles a liquid in that it has no internal boundaries and so allows light to pass through without being deflected. The difficulty in making glass is that it requires very high temperatures, such as 1600 °C to melt sand – too high for ancient furnaces to reach. Melting existing lumps of glass is much easier so what glassmakers of old did was reheat glass and when it had melted they stirred in more of the ingredients to make more glass. This sounds a bit like a chicken-and-egg situation in that before you could make glass you had to have glass. (The answer is first to melt a mixture with less sand and more of the other ingredients and then add more sand.)

As glass becomes the architectural material of the future it is reassuring to know that it will not only be there to let in light and generate electricity, but it will also control the amount of heat it lets in to the building during the day and the amount that can be lost from the building at night. A 0.3 micron film of tin dioxide (SnO_2), plus a little fluoride, will let visible light through but reflect infrared heat back and this kind of coating is applied to the glass surface by a process known as CVD (short for chemical vapour deposition). This is done as part of the manufacturing process and it is the layer on the inside of sealed glazing units. Many buildings lose most of their internal heat through the windows. At night a pane of glass allows heat in a room to pass out, although with sealed double glazing this is cut by half and with a tin dioxide coating on the glass it can be reduced to a third. A coating of silver oxide is even better and will reduce heat loss even further. The difficulty with silver is that is has to be put on the glass in a separate process which makes it more expensive, and the film is more apt to peel off.

Chemists at University College London have developed an intelligent coating for window glass that can switch from absorbing the sun's infrared (heat) rays, when rooms are cold, to reflecting those rays when the day gets too hot. The new coating is made from vanadium dioxide (VO_2) and it does not interfere with the passage of visible light. The coating is applied by CVD methods using vanadium tetrachloride (VCl_4) reacting with steam at 500 °C to form the layer of VO_2. Normally such a layer would switch from absorbing heat to reflecting heat at 70 °C, too high to be of use, but add a little tungsten and this transition is reduced to a more useful 29 °C. The only trouble with this type of glass is that it is tinted yellow which makes it impracticable at present.

Smart windows rely on liquid crystals which respond to the effect of an electric field. Under its influence such molecules will then align themselves all in the same direction and so not impede the passage of light. Remove that field and they relax to a random arrangement and the glass becomes opaque. To make smart windows requires two panes of glass, suitably coated with indium tin oxide to make the surfaces conducting. Between them is sandwiched a solution of the liquid crystals as a film only 20 microns[58)] thick. When no electric field is applied then the glass is cloudy which would be the fail-safe state if power were cut off. It was suggested that the doors of public toilets could be made of smart glass which would allow them to be seen as vacant even if the door was closed. The current would be switched off when the cubicle was locked thereby hiding the user from public gaze. For some inexplicable reason the idea has not caught on.

The trouble with glass is that it seems to attract dirt, and as we know from our own homes it has to be cleaned, especially in urban environments where city dust and vehicle fumes can settle as a thin film.

58) A micron is a thousandth of a millimetre.

The chore of cleaning windows is destined to become a less frequent activity, especially in tall buildings with glass façades when these are clad in Activ glass, produced by Pilkington and launched in 2002. This is glass whose molten surface has been exposed to CVD using titanium tetrachloride ($TiCl_4$) and steam to form a layer of titanium dioxide (TiO_2) on its surface. This layer is a mere 50 nanometres thick and is permanently bonded to it.

So how does it work? The answer is that it absorbs energy from the sun to activate an electron on the surface of the titanium dioxide which then attaches itself to a molecule of oxygen from the air to create the superoxide free radical $O_2^{\bullet}$ (the $^{\bullet}$ stands for an extra electron). This radical is capable of oxidising almost anything, not least of which is any organic dirt that is on the surface, turning it to carbon dioxide which floats away. But what of the inorganic dirt? Then it is another property of titanium dioxide which comes into play and that is its attraction for water, also brought about by the action of the sun's rays. Having removed an electron from the surface so that it becomes positively charged, it then attracts the oxygen atoms of water molecules. These in turn attract other water molecules so that there is a film of them covering the surface. Ordinary glass tends to reject water, and we can see this because rain forms droplets of water which dry out leaving spots on the glass. On titanium dioxide coated glass water 'sheets' off the glass. The result is that when rain hits Activ glass it washes its surface clean again. The discovery of Activ glass won its development team, headed by Kevin Sanderson, the coveted 'Award of Excellence' of the Worshipful Company of Glass Sellers, the oldest trade body of the glass industry and founded in the 1630s.

A new kind of self-cleaning window was launched by Henkel in 2004 and it made use of silica nanoparticles which align themselves on the glass to form an invisible film with a negative charge which also makes the glass attractive to water, in this case to the hydrogen atoms of water molecules, but the result is the same: the surface is covered with a film of water and not droplets.

Glass may be the cladding material of the future, thanks in no small part to titanium, but this element may well have an enchanting future in store.

The Magic of Titanium

"Ill met by moonlight, proud Titania" says Oberon King of the Fairies to his Queen in Act II Scene I of Shakespeare's A Midsummer-Night's Dream. What follows is a well-loved play about the revels of the night as humans and spirits cavort in a wood near Athens and Titania falls in love with an ass ...

Strange then that the more usual name for titanium dioxide is also titania, and like its namesake in the play it too has a rather mystical side to its nature as we shall see. There are abundant ores that could supply us with titanium – if we could find an easy way to free its spirit. We can obtain the metal but this is very difficult to do.

Titanium is the ninth most abundant element of the Earth's crust, and the seventh most abundant metal.[59] It has great potential as an engineering metal. Titanium melts at 1660 °C and has a density of 4.5 kg per litre, which can be compared to the other common metals such as aluminium which melts at 661 °C and has a density of 2.7, and iron which melts at 1535 °C and has a density of 7.9. There is a growing likelihood that titanium will replace these metals in some of their uses this century. It also has an oxide that has almost magical properties when bonded to glass as we have seen. Titanium dioxide is also being used to coat paving blocks, at least in Japan where the Mitsubishi Materials Corporation produces them under the Noxer brand. As this name suggests they are there to mop up NO_x fumes, the pollutant nitrogen oxide gases emitted by car exhausts. The cement blocks have a surface of titanium dioxide 5 mm thick and under the influence of sunlight they generate superoxide which reacts with the NO_x to form nitrate ions. These are then held on the surface of the paving stones until washed away by rain, or absorbed into the paving itself. Noxer blocks were originally tested back in 1997 and they have been used to replace ordinary paving stones in 30 towns in Japan.

In the 1950s, surgeons discovered that titanium metal was not rejected by the body and so was ideal for mending broken bones. It has been used in operations for hip and knee replacements, inserting cranial plates for skull fractures, and even for attaching teeth, some of

59) The most abundant elements are oxygen and silicon, after which come the metals aluminium, iron, calcium, sodium, magnesium, potassium and titanium.

which have stayed in place for up to 30 years.[60] Titanium devices for surgery are cleaned in a high temperature plasma arc which strips off the surface atoms and exposes a fresh layer of the metal. This quickly oxidises and it is this oxide film which makes it compatible with body tissue – for reasons that are still not understood.

The first titanium mineral was discovered in 1791 by the Reverend William Gregor (1761–1816) who was the vicar of Creed in the west of England, and while he knew it contained a previously unknown metal element he was unable to isolate it as a pure substance. Indeed others who tried also found it impossible to extract titanium from its oxide ore by the usual method of heating with carbon, and while this does react to remove the oxygen, the titanium then goes on to react with more carbon to form titanium carbide which is very intractable. Even when researchers had succeeded in preventing this forming they still ended up with was another intractable compound, titanium nitride. Some did manage to produce tiny amounts of the metal but their samples also contained carbide or nitride impurities which made the metal appear brittle and unworkable.

And so things remained until 1910 when M. A. Hunter, based at Rensselaer Polytechnic Institute, Troy, New York, collaborated with General Electric and obtained pure titanium by heating titanium tetrachloride and sodium metal under high pressure in a sealed vessel. His sample was 99.8% pure and he believed its high melting point would make it suitable for the filaments of electric light bulbs. However, it proved unsuitable for this purpose and the research was abandoned, but not before Hunter was able to show that it had some notable features in that it was easily worked, incredibly strong, and highly resistant to corrosion, even at high temperatures. Nor is it affected by seawater. Titanium metal can cope with all kinds of extreme conditions due to the impervious layer of titanium dioxide which immediately forms on the surface of the metal. This layer is only 1–2 nm thick to begin with, although it continues to grow slowly, reaching 25 nm after about four years. This protective layer can resist almost all forms of chemical attack and even when the surface layer is damaged it quickly repairs itself. If the protective oxide film is artificially enhanced, by anodic oxidation, it produces an iridescent surface and is then suitable for jewellery, particularly earrings.

60) Prince Charles had his broken elbow repaired with a titanium support.

The chief mined ore of titanium is ilmenite (iron titanium oxide, $FeTiO_3$) and it occurs as vast deposits of sand in Western Australia, Canada and the Ukraine. Large deposits of rutile (titanium dioxide, TiO_2) are known in North America, and South Africa. World production of the metal itself is around 90,000 tonnes per year, small compared to titanium dioxide production which is 4.3 million tonnes per year. Reserves of titanium amount to more than 600 million tonnes and while there is an abundance of this element it is extremely costly because it has to be extracted by a complicated process, and yet it could be so much more useful if it was cheaply available.

Titanium metal is obtained commercially by reacting titanium tetrachloride ($TiCl_4$) and magnesium metal at 1300 °C. Titanium tetrachloride is a crystal clear, volatile liquid which boils at 136 °C and is consequently easy to purify. The process was first shown to be viable by William Kroll in Luxembourg in 1932, although he used calcium metal as the reducing agent. He could only make it in small batches but by 1938 he had 20 kg of titanium metal. At the start of World War II in 1939, Kroll emigrated to the US and got a job with Union Carbide and later with the US Bureau of Mines. By then he had discovered that magnesium could replace calcium as the agent for releasing titanium from $TiCl_4$ and the chemical reaction generated enough heat to keep the process going. (The slag of molten magnesium chloride by-product is tapped from the bottom of the reactor and is recycled electrolytically to magnesium metal and chlorine gas, which goes to making more titanium tetrachloride.) The US Air Force became interested in titanium in 1946 and by 1950 its need for titanium alloy ensured the development of titanium refining in that country. Similar ventures were started in Russia, Japan, and the UK. As more titanium became available more uses were found for it, such as in the chemical industry, in power generation, and in surgery.

Power plant condensers worldwide contain millions of metres of titanium piping and it is claimed that none has ever failed due to corrosion. Titanium is as strong as steel but 45% lighter and it is used for lightweight alloys for the aerospace industries. It has the added bonus of being immune to metal fatigue. The fan blades of an aircraft engine are likely to be made of an alloy consisting of 90 parts titanium to 6 parts of aluminium and 4 parts of vanadium.[61)] The thin oxide layer

61) The engines of a Boeing 747 jet aircraft contain more than four and a half tonnes of titanium.

on the surface of titanium enables it to resist the corroding action of seawater and so it is used in off-shore oil rigs, and some submarines have titanium hulls. The submarine that went down to the bottom of the Atlantic to find the remains of the Titanic has a titanium hull. Some racing yachts have been made with titanium hulls. Titanium is used for propeller shafts, rigging, fire pumps, heat exchangers, and piping. Titanium metal is ideal for heat transfer applications in which the coolant is sea water or even polluted waters. Sea water is particularly corrosive to metals, which is why desalination plants rely particularly on titanium components. Titanium pipe is used in oil exploration and its light weight and flexibility make it the preferred material for deep sea exploration. If we eventually wish to tap the waves for energy then this will require vast amounts of titanium.

The Kroll process uses ilmenite as the mineral and this is heated at 900 °C with chlorine gas and carbon to form $TiCl_4$.[62] It is then sealed in a steel furnace with magnesium ingots and all the air is expelled using argon gas and the reactor is welded tight. On heating to 900 °C for two to three days the magnesium reacts to form magnesium chloride and the titanium metal forms a sponge-like solid. This is ground up, treated with strong acid to remove impurities, and then melted under argon and cast into ingots. All-in-all it takes about two weeks to convert ilmenite to titanium metal, which is why it costs as much as £30,000 per tonne. (Russia produces 40% of the world's titanium, while Japan produces 40%, and the USA 20%.)

A newer method of winning titanium from its ore has been developed and that uses an electrical process in which titanium oxide is reduced to the metal in a bath of molten calcium chloride at 950 °C on application of a 3 volt potential. This work, carried out at Cambridge University, England, led to the setting up of a company in 1998. This fledgling company has struggled to get aloft. The researchers who made it possible were Derek Fray, Tom Farthing, and George Chen and it was they who discovered that titanium dioxide really could be reduced to the metal by means of an electric current. The cathode of the electrolytic cell is the container, the electrolyte is molten calcium chloride, and the anode is carbon. The electric current converts the titanium ions to titanium metal which collects at the cathode while the oxygen ions move to the anode and are there released as oxygen gas.

62) The carbon carries off the oxygens as CO_2.

The FFC process, named after its discoverers, takes only 24 hours to make the same amount of titanium metal that takes the Kroll method a week or more to produce. It could increase the output of titanium from 60,000 tonnes a year to a million tonnes. The method can also produce titanium alloys.

The Cambridge team published an account of the process in the journal *Nature* in December 2000. There it was spotted by the US Office of Naval Research who gave them financial support and in conjunction with the British Defence Advanced Research Projects Agency they made more funding available in 2002 to allow British Titanium, as the spin-off company was now called, to set up a pilot plant. Then they began to discover that the chemistry of the process was more complex than originally realised. The company was struggling, but a third fairy godmother came on the scene in the form of Norsk Hydro, and British Titanium gave birth to Norsk Titanium in June 2005 and the research moved to Norway ... and that's where we must leave the story. Maybe one day titanium will be as plentiful as aluminium – although it may still need Titania to wave her magic wand.

Architects are beginning to use titanium cladding for buildings, an example of which is the Guggenheim Museum. This occupies a waterfront site in Bilbao, Spain, and is sheathed in 33,000 square metres of pure titanium sheet which is guaranteed to resist corrosion for more than 100 years. The roof of the new central station of Hong Kong's railway consists of 6500 square metres of titanium sheeting.

In addition to titanium metal there are four titanium derivatives which promise great benefits: titanium dioxide, titanium carbide, titanium-nickel alloy, and titanium nitride.

Titanium dioxide is the modern equivalent of white lead, but much safer because titanium compounds are not poisonous. More than half the titanium dioxide manufactured goes into paints, a quarter goes into plastics, and the rest into paper, fibres, ceramics, enamels, food colouring, printing inks, and laminates. The titanium oxide industry started in the 1930s when paint manufacturers were seeking a replacement for lead white and found titanium dioxide to have excellent covering properties. It is non-toxic, does not discolour, and has a very high refractive index,[63] which explains the brilliant whiteness titanium dioxide imparts to domestic appliances.

63) Refractive index measures a substance's ability to scatter light. Titanium dioxide's refractive index of 2.7 is even greater than that of diamond (2.4).

Titanium dioxide can be made by dissolving titanium ore in sulfuric acid and precipitating the wet oxide which is then heated to 1000 °C. The more modern process converts it to titanium tetrachloride which is then oxidised with oxygen at 1000 °C. The titanium dioxide is milled to form crystals of the right size and sometimes they are coated with aluminium oxide to make them mix with liquids more easily. A new form of titanium dioxide was reported in 2001 by a group at the Institute of Earth Sciences of Uppsala University in Sweden led by Leonid Dubrovinsky. It is one of the hardest materials ever made and maybe second only to diamond. It was made from rutile in a special anvil cell where it could be subjected to pressures of 60 GPa (600,000 times atmospheric pressure) and heated with a laser to 1500 °C. What these extreme conditions did was force the titanium and oxygen atoms into a tighter crystal array. This conferred upon titania a strength that verges on the supernatural.

Titanium carbide, TiC, is made by the action of carbon black on titanium dioxide at 2000 °C. It is the most important hard metallic material after tungsten carbide, and in fact is the hardest of all the metal carbides with a hardness rating of 9 on the Mohs scale – diamond is 10. In itself it is too brittle to be used pure but when mixed with the carbides of tungsten, tantalum and niobium it delivers great strength.

Another magical titanium material is the alloy nitinol, which can 'remember' a previous shape and return to it. Nitinol consists of 55% nickel and 45% titanium, a combination which corresponds to one atom of nickel for each atom of titanium. This alloy was developed in the US in the 1960s at the Nickel Titanium Naval Ordnance Laboratory which gave rise to the name Ni-Ti-NOL. This alloy is best known as spectacle frames which can be twisted in a way that would be permanently deformed were they to be made of any other metal but, because they are made of nitinol, they will jump back to their original shape when the pressure is removed.

Titanium will burn in nitrogen gas to form titanium nitride, a hard and corrosion-resistant material which is highly electrically conducting. Cutting tools can be made more wear-resistant when coated with a thin film of it. A coating of titanium nitride appears gold in colour and it reflects ultraviolet light. Wonderful as this material is, it is difficult to apply as a coating without using a high temperature or a very high voltage (3000 volts).

The Future's Blue

There are tens of millions of traffic signals around the world working day and night consuming electricity. Happily they now consume less that they used to thanks to light emitting diodes (LEDs), but until this century their green lights relied on old-fashioned and inefficient light bulbs because there were no LEDs that emitted light of the right wavelength. LEDs emitting long wavelength light such as red have been known for many years but ones that emitted shorter wavelength light, such as blue, had proved impossible to find – that is until gallium nitride (GaN) came along. Sales of devices containing GaN are now in excess of $2 billion a year worldwide. Indeed gallium nitride is the new semiconductor that promises to outshine silicon and gallium arsenide, and revolutionise many areas of life, quite apart from reducing the demand for electricity.

Every time we use a mobile phone, rely on satellite navigation, play with a game-boy, listen to our iPod, log on to the internet, or work on our laptop, we rely on semiconductors. Within a few years these little workhorses will be able to monitor our car to maximise its efficiency, identify us by our fingerprint or irises, and diagnose our illnesses. At the heart of all this new technology will be III/V semiconductors, of which GaN is the rising star. The Roman numerals III and V refer to an old form of the periodic table in which elements like gallium and indium were in group III (now called group 13) and elements like nitrogen, phosphorus, arsenic and antimony were in group V (now group 15). There are several kinds of III/V semiconductors, the best known of which are gallium arsenide (GaAs), gallium phosphide (GaP), indium phosphide (InP), and indium arsenide (InAs). All are used in light displays. GaAs and InP are also used for high frequency generation and amplification, and GaAs, InAs, and InSb are used to make laser diodes.

Scientists began to take an interest in gallium arsenide as a semiconductor material because it generates less heat than silicon and is therefore more suitable for use in supercomputers as well as in the more mundane mobile phone. Silicon and GaAs semiconductors are still the most used, but they have their limitations. GaN semiconductors are ten times more powerful and can carry much more current. They are also a lot more robust. So what can this new material be used for? The most obvious use is to provide a new generation of LEDs, ones that emit blue and green light, and they can generate white light when combined with those which emit at the red end of the spectrum. The emission of green light will widen its use not only to traffic lights but to all kinds of illuminated displays, while white light gives it potential to be a source of general indoor illumination. GaN also makes blue lasers possible and they in their turn will increase the amount of data that can be stored on a CD or DVD; in theory a disk the size of a thumb nail could hold 50 gigabytes of information. All these uses come as a result of GaN's large bandgap – see box – and it is this which gives it the ability to provide light at the short wavelength end of the spectrum.

Bandgap

If any material is to be useful as a semiconductor then the energy needed to promote an electron from the valence band to the conduction band has to be just right. The valence band is where the electrons that bond atoms together are located but there are regions of space to where they can move freely and this is the conduction band. How much energy is required to boost an electron to this band determines whether the material is a conductor, a semiconductor, or an insulator. If the energy gap is too large to cross then it is an insulator, such as diamond which has an energy gap of 5.3 eV,[64] if the energy gap is small and easily jumped then it is a conductor, and for many metals the gap is zero or less than 0.1 eV. For material with an energy gap of between 0.1 and 3 eV then it can be bridged but needs more energy and it is classed as a semiconductor. Silicon has a band gap of 1.1 eV and germanium 0.69 eV making them both potentially useful semiconductors for microelectronic applications. In gallium arsenide it is 1.4 eV, which is ideal for optical-electronic devices such as CDs and its light is red. With GaN the gap widens to around 3 eV, which is just bridgeable, and it emits light at the high energy, short wavelength, end of the spectrum and it is this that has opened up a whole new area of applications.

64) This is a unit of energy used in electronics and is the energy an electron acquires when it is accelerated across a potential difference of 1 volt.

In the mid-1990s Shuji Nakamura, working for the Nichia company in Japan, developed the first LED capable of emitting intense blue light, and in so doing he completed the range of colours of the visible spectrum so that all colour combinations now became possible. He had discovered the wonders of GaN which resulted in the blue laser diodes of Blu-ray disc technology which reach us in the form of things like the Sony PlayStation 3.

Because GaN can tolerate high voltages without losing its semiconductor properties it will have application in power transmission switching gear. Because it has the ability to withstand high temperatures it could find use in monitoring the performance of engines and other devices. Because it can carry a lot of current it will find use in wireless base stations for radio frequency transmissions. The cellular phone industry is now text-based rather than voice-based, as evidenced by the popularity of the third generation devices. These demand greater signal purity from the power amplifiers and this is what GaN delivers, and the base stations for mobile phones could be ten times further apart than at present. Because it can operate at high frequency GaN will be used in micro-mechanical integrated circuits for wireless broadband access. And because it displays the piezoelectric effect, in other words when it is squeezed it emits electricity, it will be used in pressure sensors.

GaN is an attractive material and it retains its semiconductor properties up to 1000 °C, well above temperatures at which silicon fails as a semiconductor. GaN can withstand more than seven times the electric field that causes GaAs to undergo ionization and so end its semiconductive ability. If voltage and current are considered together, then GaN can deliver a 50-fold better power performance than GaAs. GaN has attracted attention from weapons manufacturers, where expense is of minor importance, because GaN offers more compact systems for electronic warfare, such as phased array radar. This has a longer range, does not reveal its location because it has no rotating scanner, and would even continue to work should it suffer a burst of intense radiation following a nuclear weapon attack because GaN is not as sensitive to such ionizing radiation as other semiconductors.

A domestic of GaN is in radiofrequency transistors of the type we rely on in microwave ovens where currently magnetrons are used to generate the heating rays.

The puzzling thing about GaN is that it shouldn't work as a microelectronic material: it is just too riddled with crystal dislocations. If GaAs has more than 1000 dislocations per square centimetre it is unable to function as an LED, yet GaN can have a *billion* dislocations per square centimetre and still work. How does GaN accomplish this mission impossible? As yet there is no clear answer to this question although it has something to do with the interface between the p- and n-forms of the semiconductor. The p-GaN is obtained by adding traces of electron-deficient atoms such as magnesium to create the positive 'holes' while n-GaN is achieved by doping with silicon. Between them is sandwiched the semiconductor indium-gallium nitride (InGaN) which will also operate with a high level of lattice defects. This is known as the quantum well region into which the 'holes' and electrons migrate under the influence of an electric potential, there to combine to form what is called an exciton which releases a photon of light. Normally the regions where this happens have to be free from crystal defects but GaN can tolerate such defects and still allow excitons to form.

Production of crystalline semiconductor GaN has not been easy, mainly because there is no perfect surface on which to grow it by CVD. Silicon substrates for GaN were researched in the 1990s but this approach was abandoned because the structure of silicon did not really match that of GaN. Sapphire is an alternative substrate but this material suffers from low heat conductivity, although it has the advantage of being transparent and cheap. Another surface which has been used is silicon carbide, which has good thermal conductivity, and this is often the surface of choice. Of course the ideal material on which to grow GaN would be GaN itself and Sumitomo Electric Industries Ltd of Japan has developed single-crystal GaN for this purpose and is currently producing them at a rate of 500 units per week, each of which can be used to make 10,000 blue laser diodes which are being used in the Blu-ray Disc system.

What's So Special about Gallium?

Gallium was discovered in 1875 by Paul-Émile Lecoq de Boisbaudran at Paris, France, when he examined the spectrum of a zinc sulfide ore from the Pyrenees and saw a faint blue-violet line which told him a new element was present. Its existence had already been predicted six years earlier by the Russian chemist Dimitri Mendeleyev, the man who drew up the first periodic table and saw there was a missing element below aluminium in his group III.

Gallium is more abundant than lead in the Earth's crust so it is unlikely there would ever be a shortage of this metal, but it differs from lead in that no gallium ores as such exist and it has to be extracted from other sources. It is present in small amounts in several ores such as the aluminium ore bauxite which is the source of the 200–300 tonnes of gallium that is produced annually. Pure gallium cost around $3000 per kg in 2006 and for semiconductors it has to be 99.99999% pure, in other words to have only 1 foreign atom for ever ten million gallium atoms.

Gallium is a silvery-white metal which is soft enough to be cut with a knife. It becomes liquid if held in the palm of the hand because its melting point is 30 °C, and it shrinks in volume as it does so which is rare because most substances expand on melting.[65] It boils at 2403 °C making it the substance with the longest liquid range.

65) Ice also shrinks on melting, as does the metal antimony.

The future of GaN could well be in the form of nanowires which might one day be used for nanoscale electronics, as in biochemical sensing devices. (Unlike semiconductor GaN these are free of defects.) These were first produced by Peidong Yang of the Lawrence Berkeley National Laboratory in California and he grew them on zinc oxide nanowires by exposing these to a mixture of trimethyl gallium and ammonia at 700 °C whereupon the molecules decomposed and the gallium and nitrogen atoms combined on the surface of the zinc oxide to form a sheath of GaN. The zinc oxide was then removed by chemical means leaving hollow tubes of GaN nanowires. Nanotubes are currently a hot topic of research as chemists struggle to devise nano machines that can operate at the level of the atomic world. Meanwhile GaN will continue to provide us with light from the short wavelength end of the spectrum in devices that require less and less electricity to make them work.

This chapter has shown how it is possible to produce materials that will allow us to continue enjoying the benefits of the present in a future that will have to provide them with less and less reliance on non-renewable sources of energy. The answer will come partly from the rarer metals of this planet, although one of those may well be the limiting factor to all that we might achieve, and that metal is indium.

Issue: Is the World Running Out of the Rare Metals We Need to Make Solar Panels?

There is a small cloud on the horizon of solar energy generation using the high efficiency gallium indium phosphide, and that is its dependence on indium. Indium is also crucial to many other solar cells because it is used as indium tin oxide to provide a transparent electricity-conducting coating on the surface of glass. Indium will 'glue' itself to glass in a way no other metal does, which is why it is so widely used. Indium comprises only 0.1 ppm of the Earth's crust, making it the 69th most abundant element, in other words it is extremely rare. There are no indium mines nor ever likely to be.

The world's supply of indium comes as a by-product of zinc and lead refining and is produced in China (110 tonnes per year), Japan (70 tonnes), Canada (50 tonnes), Belgium (40 tonnes) France (10 tonnes), and Germany (10 tonnes), with smaller production in a few other countries. The total refined indium production is around 340 tonnes per year, most of which goes into indium tin oxide (70%) and indium semiconductors (15%) with a few other uses, such as low melting alloys in fire-sprinkler systems in shops and warehouses, accounting for the remainder,.

As a semiconductor, indium has to be at least 6N grade, which means six nines, i.e. 99.9999% pure, but some uses require it to be 7N. Only a few producers can achieve this purity, such as the Indium Corporation of America, PPM in Germany, MCP in the UK, and Japan Energy. The price of technical grade indium (3N) was around $100 per kg until early 2003 when it doubled in price as more uses for it were discovered. The indium for solar cells has to be 6N grade and this costs more than $2,400 per kg.

How much indium can be extracted from zinc and lead ores remains to be seen but there will never be enough to generate all the electricity we would like to extract from solar cells by the end of the century. What the world needs now are some bright young chemists to find alternative substances. One group of strong contenders will be carbon nanotubes and these are already finding application in LEDs and PV solar cells. Meanwhile there are 30 or so minor metals and thousands of their compounds waiting to be explored. Maybe one of them could even outperform indium.

6

Better Living (IV): Chemistry in the Home

A small arrow printed before a word in the main text indicates that there is more information on that topic in the Glossary.

News from the Future

Global Times News, 21 March 2025

Laundry Tablets Go Green All Over

Boxes containing a year's supply (100) of the new laundry capsules, Green Clean, will soon be on sale in supermarkets across India and are made entirely from renewable resources. Research chemists at IndiaChem, Bangalore, have finally solved the problem of making all the ingredients this way.

"The most difficult challenge we faced was finding a chemical that would neutralise the calcium of hard water" said Donald Patel, their head chemist, "but in the end we returned to using sodium tripolyphosphate which can be manufactured from the phosphate extracted from sewage works." Dr Patel pointed out that although this chemical was once accused of causing river and lake pollution, its removal from waste waters no longer made it a threat and it is recycled as raw material for the chemical industry.

The other ingredients, the surfactants, the dirt solubilisers, the dye stabilisers, the whitening agents, the fabric softeners, and even the bleaches are now made from chemicals obtained from crops, including straw and bark, or from the minerals obtained from sea water.

"Green Clean works at room temperature, and uses a minimum volume of water" added Patel, pointing out that last year saw the launch of a new washing machine for India in which the rinse water from one wash is disinfected and stored to be used for the main wash next time. "We can now boast that washing clothes is as environmentally friendly as it is possible to be" he said.

The various ingredients in the new detergent are microencapsulated in coatings which ensure that they release their chemicals at the right time, and the whole 25 gram of detergent is sealed in a water-soluble polymer, polyvinyl alcohol, which is also produced from plant materials.

Page 7: Guru says tablets are unclean and cause leprosy.

If humans had evolved to be fur-covered animals there would be no need for clothes, but then we would have been denied many of life's little pleasures that are associated with them. Yet down the centuries

Better Looking, Better Living, Better Loving. John Emsley

ISBN 978-3-527-31863-6

women especially have paid dearly for our need to wear clothes, in terms of the effort they have expended in washing them. In developed countries that burden has been lightened, but in some respects it has become a burden on the environment, because in order to clean clothes in the modern home we require energy and chemicals. In this chapter we will look at the way the latter have evolved to give us products that are not only much better than they used to be, but are more environmentally friendly. We are already well on the way to a future in which the forgoing news item could well come true, maybe even before 2025.

The first topic focuses on the seemingly mundane world of detergents, and shows how chemists have managed to keep abreast of all the changes that consumers, the environment, and even the law, have demanded of them, and produced products that earlier generations would have marvelled at. Whether future generations will be able to take advantage of them will require the chemical industry to be based entirely on sustainable resources and this is an issue we will consider at the end of the chapter. Before then we will discover how surfactants save the lives of premature babies, how we can wash clothes clean in cold water, how an incorrect news report changed a nation's way of doing its laundry, why lots of lather is not always the best way to get things clean, and how unpleasant smells can be dealt with. Chemistry and cleaner living should go hand in hand.

Wash All My Troubles Away: Laundry Aids

The traditional method of washing clothes was to go down to the nearest river, immerse the clothes in the water, then beat them against smooth rocks until they were clean. Some women in the world still wash clothes this way, while many more rely on the beating action of the revolving drum in a front-loading washing machine, but it is mainly the detergent in the water which now removes the dirt. Those with top-loading machines rely entirely on the dirt-removing properties of detergents and bleaches but their way of washing uses much greater volumes of water.

In the past 50 years lifestyles have undergone many changes, not least in the kinds of garments that we wear, and the materials from which they are made. For environmental reasons we want to use less water, less energy, and less detergent, yet we now wash clothes much more frequently, sometimes merely to refresh them rather than remove dirt. We have other wishes as well. We would like to wash new clothes without the dyes running; we don't want clothes to look wash-worn when they have been laundered many times; we want them to feel soft to the skin after washing and drying; and we want the detergent to be gentle to delicate fabrics. If we stain our clothes we want our detergent to remove even the most difficult of stains such as curry sauce and coffee, mud and motor oil, or grass and grease, as well as the 101 ordinary stains of everyday life.[66] All these demands have been met by modern detergents.

Today detergents come in a variety of formats, such as liquids, powders, sachets, and tablets and they may have as many as 30 ingredients.[67] If you don't like the idea of so many chemicals being needed to meet all the above requirements then there are novelty products such as ceramic disks that you can put in your washing machine and which are supposed to increase the beating action so you need add much less detergent. Laboratory tests showed that they were virtually useless. On the other hand a premium brand laundry product should now ensure that all stains are removed, that clothes are not damaged, the washing machine is not corroded, and that there is minimum impact on the environment by using less water, less energy, and less detergent. Indeed it is now possible to wash clothes clean in water as cool as 20 °C, something that our great grandmothers would not have thought possible because they needed to *boil* the laundry to ensure all stains were removed. For most people, doing the washing is now as effortless as putting things in the machine, priming the dispenser with detergent and fabric softener, choosing a wash cycle, and an hour or so later the job is done. Clothes are now washed so frequently that in many families the washing machine is in use every day.[68] Wear once then wash is the order of the day.

Our early ancestors had quite a struggle to wash clothes, but they were willing to experiment. The first laundry aid they discovered was a natural chemical, saponin, which is found in the leaves and roots of

66) Surveys have shown that the most common stains on clothes are mud, oil, coffee, and wine.

67) European consumers prefer powders and tablets, US consumers prefer liquids.

68) A survey conducted by Unilever in 1997 found that the average family does about 5 laundry washes per week.

the soapwort plant *Saponaria officinalis*. This is a complex molecule which combines steroids and carbohydrates and it forms a rich lather in water which could be used to assist washing. (It would never be permitted to be used in modern detergents because saponins are slightly toxic and attack red blood cells.) Soapwort, which has pretty pink flowers, grows naturally by the side of rivers and a solution made by boiling its leaves and roots was especially useful for washing wool. But why would any plant need to make its own detergent? The answer is to protect itself. Research in 1998 at the John Innes Centre in Norwich, England, showed that when oats, which also produce small amounts of saponin, were genetically modified to remove the saponin-producing gene they lost their natural resistance to fungal diseases which tend to flourish in the damp places and climates where such saponin-producing plants grow.

Better than soapwort is soap itself which first appeared in the Middle East around 4500 years ago, although it was not then regarded as a laundry aid but as a toiletry. The rich of the ancient world employed fullers to clean their clothes and they worked by pounding the washing with their feet in a tub of warm water made alkaline with a little natural soda (sodium carbonate). Indeed the hieroglyph for a fuller in ancient Egypt was a pair of legs standing in water. Soap was used by the Romans and the author Pliny writes of *sapo* which was made from goat tallow. It came in two varieties: a solid kind which was made using soda (sodium carbonate) and a liquid type which was made using wood ash (potassium carbonate).

About a thousand years ago, soap began to be manufactured and used as a laundry product. The first manufacturer in England set up shop in Sopar's Lane in London in 1259. However, it was the discovery of a way of manufacturing soda from salt on an industrial scale in the late 1700s that turned the production of soap into a major industry, producing toilet soaps for personal use and household soaps for laundry and general cleaning. Clothes may have got cleaner but the family wash was still a weekly grind of soaking, scrubbing, boiling, pounding, wringing, rinsing, wringing and drying.

What is Soap?

Soap comes from fats and oils. These natural chemicals are composed of glycerol (chemical formula $C_3H_8O_3$) to which is attached three long chain fatty acids, producing a molecule which is rather like a capital letter E with elongated horizontal strokes. The vertical upright of the E represents the three carbon atoms of the glycerol molecule and the horizontal strokes are fatty acids, of which the best for making soap are the saturated kind and these are the most common ones, namely palmitic acid, with a chain of 16 carbon atoms, and stearic acid, with a chain of 18. When fats and oils are heated with an alkali like sodium carbonate, the fatty acids are freed from the glycerol and form sodium palmitate and sodium stearate which is what soap mainly consists of. In the early days the glycerol was left in the soap, but today it is separated and recovered and used as a natural resource for the chemical industry.

Real progress in relieving some of the washday chores came in 1907 with the introduction in Germany of Persil, the first modern detergent. Its ingredients were soap, sodium carbonate, sodium silicate, and sodium perborate, and when clothes were boiled in water containing it they came out of the wash perfectly clean. Things got a little better in 1913 when a pre-wash product became available, also in Germany. This was based on protease enzymes which are capable of digesting protein stains such as blood, egg, meat juices, and milk. The market for such laundry aids expanded in the late 1950s when better enzymes became available, but they were still regarded as pre-wash agents. In the 1970s they were joined by amylase enzymes which will break down starch stains into soluble components. Then in the 1980s along came lipase enzymes which will digest fats and oils, and these comprise some of the more difficult-to-remove stains such as olive oil, cooking oils, butter, margarine, cream, and sebum. (Sebum is a mixture of dead skin cells and oil and it stains clothes which come into close contact with the body, such as around cuffs and collars.)

Enzymes have improved over the years so that food stains on clothes, and unmentionable stains on undergarments, now present no problem. In the late 1980s, enzymes were engineered that could resist the action of the peroxide bleach and by 2000 it was possible to have a mix of all kinds of enzymes in the same detergent. Because enzymes are themselves protein molecules, a *protease* enzyme could in theory attack other enzymes and digest them as well, so special varieties have been developed to overcome this. Today, enzyme demand is in excess of 30,000 tonnes in Europe alone and most of this ends up in detergents. Not that this has prevented a rather curious delusion about enzymes persisting in the UK where many people deliberately eschew their benefits – see box.

The Curious Tale of the 'Non-biological' Detergents

About 25 years ago a BBC television programme in the UK gave air time to a group of people who complained that a new detergent (Persil New System) was causing them to break out in rashes and other skin complaints such as eczema, and they blamed its 'biological' ingredients, by which they meant the enzymes. There was no scientific evidence to support their claims – and there never has been – but they became widely believed and today 'non-biological' detergents, which contain no enzymes, have around a 30% share of the detergent market in the British Isles. Nowhere else in the world are such products produced, but the belief that enzyme residues cause skin diseases is now a deeply rooted urban myth: undeniable, undesirable, but untrue nevertheless.

In the past century detergents continued to do more and more of the washing and today all kinds of ingredients are included apart from enzymes. These are:

- *surfactants,* to lift dirt off fibres;
- *foam regulators,* to prevent a build-up of foam which can overflow from the machine;
- *builders,* to make the wash water soft by trapping calcium;
- *anti-redeposition agents,* to prevent dirt from depositing itself back on clothes;
- *dye transfer inhibitors,* to prevent dye molecules from colouring other clothes;
- *peroxide bleaches,* to remove stains such as tea, coffee, and fruit juices;
- *bleach activators,* to enable peroxide bleaches to work at lower temperatures;
- *fluorescent whitening agents,* to make whites appear whiter and to disguise the yellowing which comes with age;
- *anti-corrosion inhibitors,* to protect the metal parts of the washing machine;
- *fragrances,* to impart a pleasing smell to washed clothes, towels, and bedding.

Quite sophisticated chemistry takes place in a modern washing machine thanks to these chemicals. So how do they work their magic?

Surfactants[69]

There is nothing unnatural about surfactants. Sometimes when a man urinates into a toilet bowl he is surprised to see that it froths up as if his water contained a foaming agent. In fact it does, and it comes from his own urine discharging surplus natural surfactants. The human body produces these chemicals and they are particularly essential for the working of the lungs.

Many premature babies used to die from respiratory distress syndrome because their lungs lacked a surfactant to keep the small air spaces open. The high surface tension of water tends to keep these closed and this has to be lowered by surfactants which a premature baby may not yet be capable of producing. The human surfactant is made up of fats and proteins and, like detergent surfactants, they have the ability to reduce the surface tension of water, in this case allowing lungs to expand easily as we breathe in, but when we breathe out they become rigid and this prevents the very fine air passages in our lungs from collapsing. Premature babies are now given a modified version of the natural surfactant until they can produce their own, and this has halved their death rate in recent years.

Oil and water have to be made to mix if we want to wash grease and grime from clothes, and this is the role of the surfactant, which is a molecule with a head and a tail. The head is water-compatible, such as a salt group, while the tail prefers the company of oils and grease, and is a hydrocarbon chain. The head of most surfactant molecules attracts water by carrying a negatively charged group of atoms. In soap this is a carboxylate group; in synthetic surfactants it is a sulfate or a sulfonate group[70]; in human surfactants it is a phosphate. The tail of the surfactant has to be a hydrocarbon and there is less variability in this part of the molecule except for the length of the carbon chain which is generally between 12 and 20 carbons long.

The first and most important effect of a surfactant is to make water wetter, which sounds incongruous but what it means is to make wa-

69) The word is derived from the phrase 'surface active agent'.

70) A sulfate group is SO_4 and a sulfonate group is RSO_3 where R is an organic molecular component.

ter better able to wet fabrics and it does this by lowering its surface tension, and this is what happens as the surfactant molecules cluster along the water's surface. When there is no longer room for all the surfactant molecules at the surface they begin to form spherical aggregates, called micelles, with the water-attracting ends on the outside and all the oil-loving hydrocarbon tails on the inside. It is this internal 'oily' environment that attracts the grease associated with dirt and grime which diffuses into the micelle where it is happy to remain until it is flushed away. With the sticky grease gone, the rest of the dirt then falls off the fibres and into the water, where other molecules are waiting to disarm it as we shall see.

There are different ways of attracting water molecules, which is why there are four kinds of surfactant and these are defined by the water-seeking head.

Anionic surfactants have a negative head and the major ones are sodium alkyl benzene sulfonate, which is used industrially as a cleaner, and sodium lauryl sulfate and sodium laureth sulfate which are used in household detergents and toiletries – see Chapter 3. Soap is also an anionic surfactant. Anionic surfactants produce copious suds and are excellent at removing fats and grease.

Cationic surfactants have a positively charged head and these are used as fabric conditioners and even antiseptics – see page 168. They are attracted to fibres and will coat them with a layer one molecule thick, but that is enough to prevent strands matting together and this is why they make clothes feel softer. One that is often used is lauryl trimethyl ammonium chloride, and various hair conditioners contain about 3% of this surfactant.

Amphoteric surfactants have both a positive and a negative component at the water-attracting end. Amphoteric surfactants are used in washing-up liquids, but rarely in laundry detergents because they cost too much. The best known amphoteric surfactants are the cocamides.

Non-ionic surfactants are neither positively nor negatively charged but instead they have several oxygen atoms at the head end of the molecule, generally in the form of a series of →ethoxy groups.[71] They produce less foam than anionic surfactants and are used in low temperature laundry detergents because they work well even at 30 °C.

71) These have the chemical composition CH_2CH_2O and there can be two or more of these units at the head end.

Most surfactants that are manufactured are anionic, next come non-ionic, then cationic, with only a small proportion of amphoteric ones. In total around 9 million tonnes of surfactants are manufactured globally every year, of which soap still accounts for a quarter. Surfactants are not only used as detergents for household cleaners and toiletries but in paper manufacture, paints, plant protection and even in foods. (For example, the natural surfactant lecithin, which is an amphoteric surfactant, is used as an emulsifier in chocolate and ice cream.) Surfactants are used as wetting agents, emulsifiers, dispersants, solubilisers, surface conditioners, bactericides, viscosity modifiers, anticorrosives – and detergents. In some specialist applications the hydrocarbon 'tail' of the surfactant is a →silicone chain or even a →fluorocarbon chain. The former can be used in pesticide formulations in agriculture, the latter to create foams for fire fighting.

Foam Regulators

The lather on top of the water has psychological value when washing dishes by hand because it is used to judge how much detergent is still active, and for that reason dishwashing liquids contain non-ionic surfactants specifically to generate foam in addition to absorbing grease. Just the opposite is the requirement for front loading washing machines and dishwashers where foam is a real disadvantage and indeed foam suppressants are added to the detergent such as silicone oil, although only in tiny amounts. Somewhat surprisingly soap itself can act as an antifoaming agent when added to synthetic surfactants and what it appears to do is disrupt the uniform layers of surfactant molecules normally needed to support a bubble and so the bubble collapses.

Builders

The builders in a detergent are there to improve the quality of the wash water. The first builder was simply washing soda which dissolves readily in water and raises its →pH so that it is alkaline. This in itself has a cleaning action because some of the negative hydroxide ions which now predominate are able to react with grease and break it down into water-soluble components, namely fatty acids and glycerol. In fact the higher the pH the better the grease removal, but it can have a major drawback because it can be detrimental to the fabrics being washed.

The principal role of builders is to bind to calcium to prevent it interfering with the surfactant.[72] Ones that have been added as ingredients to detergents are sodium tripolyphosphate, citric acid, and zeolites. (All are capable of entrapping more than 100 mg of calcium per gram of additive.) It is the last one of this group which is now preferred because the environmental implications are minimal, as we might expect of something that is essentially an expanded form of sand (silicon dioxide) with cavities into which the positive calcium ions wander and then remain stuck to the negatively charged oxygen atoms which are there. Another builder is sodium silicate (also known as water glass) and this too can neutralise calcium to some extent, but it is a good corrosion inhibitor and protects the metal parts of the washing machine.

A much better calcium sequestering agent is sodium tripolyphosphate (STPP) which is also good at keeping in solution the dirt and clay that had been lifted from clothes, and STPP became the standard detergent builder in the 1950s and 60s, but it was overused. Its use coincided with the fouling of natural waters by blue-green algae, especially in lakes with large cities nearby, such as the Great Lakes on the US–Canadian border. Some of these became choked with a green slime of algae which suffocated other species under their decaying masses, and phosphate appeared to be to blame. In the 1970s and 1980s detergent manufacturers found themselves under pressure to stop adding STPP to their products and to replace it with alternative builders. They decided that zeolites would best fit the bill and these are now the main kind of builder but STPP is still needed for dishwasher detergents. There is now less clamour to ban STPP because it was finally realised in the 1990s that phosphate was only partly to blame; it was the disappearance of the zooplankton that feed on algae that allowed the latter to multiply unchecked. The zooplankton had been killed by pollutants such as heavy metal effluent from industry and pesticide runoff from farmland.

Whether your laundry needs the benefit of builders depends on the local water supply. Is it hard? If so you will need a builder. Or is it soft? In which case you won't. Almost all water in Japan (92%) is soft and

72) Dissolved magnesium ions may also be present and they can have the same detrimental effect.

there is no hard water there, whereas in the UK more than 60% of homes are supplied with hard water. In Germany hard water affects 50% of homes, in France and Spain around 45%, in Italy about 20%, but in the USA only 5%.

What is Hard Water and Why Is It So Hard to Discuss Scientifically?

Water is described as hard when it contains a lot of calcium ions. While such water may be beneficial as a drink, it is not good for washing clothes. Measuring hardness might appear relatively easy and all that is needed is to analyse the amount of calcium dissolved in the water. How this information is presented is another matter, and it varies from country to country. The amount of calcium can be calculated in terms of milligrams of calcium ions per litre of water, or milligrams of calcium carbonate per litre, which might seem the most sensible because this is the source of the calcium dissolved from rocks that the water has been in contact with.

Water hardness was important long before modern methods of measuring it were established and for this reason it is reported in rather parochial ways. Thus the English system records it as *grains* of calcium carbonate per litre, and this Medieval unit of weight is 65 mg. The American water hardness scale is based on grains of calcium carbonate per US *gallon*, which is equivalent to 3.78 litres. The German system uses milligrams of calcium *oxide* (CaO) per litre, which is not the chemical form in which it is present. The French system does take into account that it is more like dissolved calcium carbonate ($CaCO_3$) and their system is based on *moles* of calcium carbonate per cubic metre, of which a mole weighs 100 g and is calculated from the atomic weights of the component atoms (Ca + C + 3O = 40 + 12 + 48 = 100 exactly) The French system is the one preferred by chemists in industry and it produces a scale of hardness ranging from 0 to 30 or more.

Soft water has less than 90 mg of calcium carbonate per litre, medium water has between 90 and 270, and hard water has more than 270, which in terms of FH (French hardness) correspond to less than 9 for soft, 9–27 for medium to hard, and above 27 for very hard.

Anti-redeposition Agents

The traditional anti-redeposition agent is sodium carboxymethylcellulose (CMC), a soluble form of cellulose in which most of the hydroxy groups have been replace by acetate groups. CMC is present in detergents to the extent of around 0.5% of their weight but this is enough to bind to the fibres of cellulosic fabrics, such as cottons and linens, thereby making them better able to repel both the grease-containing micelles and the dirt particles. CMC also makes a good thickening agent for laundry liquids.

Dye Transfer Inhibitors

Whenever something new that is highly coloured is washed, such as a T-shirt, a lot of the dye may end up in the water. If that garment is washed along with white or light-coloured clothes then there is a real risk that they will pick up some of the dye and emerge slightly coloured and maybe regarded as unwearable as a result. At the other end of the spectrum there are black and deeply coloured clothes that begin to turn pale after many washes as the dye is leached or bleached away at each wash.

One answer to washing dyed fabrics is to separate the laundry into whites plus light-coloured fabrics, and black plus dark-coloured fabrics, and wash them separately and there are detergents for just such washes, the former containing bleaches, the latter are without them. Even so this does not entirely solve the problem of dye transfer and it was only in the 1990s that chemists came up with an answer: PVP, short for poly(*N*-vinylpyrrolidone). This polymer is water soluble and along its chain there are chemical groups that attract and hold on to those dye molecules that have floated off into the wash water, thereby preventing them from attaching themselves to other fabrics.

Peroxide Bleach and Bleach Activators

The original Persil contained sodium perborate as a bleach. This simple chemical, formed from boric acid and →hydrogen peroxide, is a white powder which is stable and water soluble, but it only begins to act as a bleach when the temperature of the water reaches 60 °C, and it works even better at temperatures higher than this. That suited the housewives of the early 1900s who still expected to wash the family's laundry in a copper boiler. Hot water washing continued up to the 1970s with all washing machines having heating elements that could ensure high temperatures. Detergents in Europe then contained between 20 and 30% perborate, although elsewhere it tended to be less than 10%. Another solid bleach is sodium percarbonate which replaced sodium perborate because the boron in the wash water passed unchanged through sewage works and was thought to be toxic to some aquatic organisms. The new bleach was made from sodium carbonate and hydrogen peroxide. [73]

The two oil crises, of 1973 and 1979, saw energy become much more expensive. It took a lot of electricity to heat the large volumes of water then commonly used to wash clothes, and people were encouraged to wash at lower temperatures, such as 40 °C, but at this temperature sodium perborate bleach is not very effective. Nor does sodium percarbonate fare much better. They both release hydrogen peroxide into the wash water but this is not sufficiently active as a bleach at these lower temperatures. The chemists' answer was to find something that would activate the hydrogen peroxide at 40 °C. Many substances were tested and found to work, and they did this by converting the hydrogen peroxide into peracetic acid. One of the best at doing this was TAED, short for tetraacetylethylenediamine, which is preferred because it has the advantage of being relatively cheap. TAED reacts with perborate and percarbonate at 40 °C or lower, and forms peracetic acid which then acts as the bleaching agent. The bleachable stains on clothes and table linen due to red wine, tea, coffee, colas, and fruit juices were once again easily removed. In the US and Japan a different activator was preferred: NOBS (short for nonanolybenzene sulfonic acid) which gives improved performance in cold water washing.

Other progress towards finding a bleach combination that would work well in cold water came with the discovery that a certain type of manganese compound was capable of catalysing the bleaching. In a paper in the June 23rd issue of *Nature* in 1994, Unilever chemists reported the new catalyst which consisted of a manganese atom surrounded by a ring molecule with three nitrogen atoms. So well did this catalyst work that Unilever added it to their detergents but, alas, it had an Achilles heel and within weeks it was dead – see box.

73) The manufacturers of hydrogen peroxide and peroxide bleaches operated an illegal cartel for several years and were fined €400 million by the European Commission in 2006.

Sometimes the Chemistry Can Be Too Good

Adding a little of a manganese compound to a detergent containing a bleaching agent makes it work much better because the manganese acts as a catalyst. The enhanced removal of difficult stains was remarkable, even at 20 °C, and extensive tests by independent laboratories, and by 60,000 consumers, confirmed it worked superbly well. So, in May 1994, Unilever launched a new detergent across Europe, calling it Persil Power in the UK, OMO Power in France and Holland, and Skip Power in Germany, and indeed it was spectacularly good and it boosted sales. But it did something strange to rayon (artificial silk) although it did not appear unduly to affect linen and cotton which are also cellulose-based fibres. Even so a rayon garment needed to be washed 20 times or more for the damage to become evident.

Unilever's main competitor, Procter & Gamble, challenged the new product with evidence from leading research organisations that the catalyst in the new detergent damaged clothes. Unilever responded by reducing the amount of manganese catalyst by 75% but this was not enough and eventually, and in response to adverse publicity, they withdrew the new detergent after only six months. It appears that the manganese catalyst released the peroxide of the bleaches in the form of free radicals and while these were excellent for attacking stains they also attacked cellulose. Why were cotton and linen fabrics not equally susceptible? The answer lay in the length of the polymer chains from which these fabrics are made. Those of rayon are shorter than those of cotton and linen with the result that broken chains quickly translated into much weaker fibres.

Fluorescing Agents

These are the ingredients which make white fabrics appear brilliantly white, and keep them looking that way. After repeated washings many such garments take on a faint yellowish tinge, mainly due to traces of iron being deposited on the fibres. In the 1800s the remedy to yellowing fabrics was to add a blue dye[74] to the final rinse water to counter this effect. The answer today is to add about 0.5% of a fluorescer to the detergent and indeed it is possible to claim that the product washes 'whiter than white' which sounds absurd but in fact contains a germ of truth. Fluorescent whitening agents are complex molecules which incorporate as part of their structure a group of atoms that can absorb ultraviolet light and then lose that extra energy

74) This was synthetic ultramarine. This strange chemical is discussed in Chapter 7.

by emitting it as light of a slightly longer wavelength, i.e. as blue light. The result is to hide any yellowness that comes with frequently washing.

Fragrances

Detergent manufacturers began to add fragrances to their products in the 1950s with the object of masking the smell of surfactant ingredients which have a greasy odour. This became more important once the washing machine was a part of the living space of the home. Fragrances also gave a fresh smell to the finished laundry and people liked this. All kinds of fragrance molecules are used to achieve a pleasing odour and they can persist through the drying and ironing stages. Some people claimed to have skin that was sensitive to detergent residues and as a result the fragrances thought to be responsible were phased out. Allergic reaction to traces of fragrances on clothes has never been scientifically proven, and indeed when consumers who have claimed to be so affected have taken part in →double-blind skin-patch tests, the results have shown that detergents were not to blame.

The need for products to wash our clothes is met by industry to the tune of around 22 million tonnes of chemicals per year. In some countries the use of laundry soap still outranks detergents, such as in Kenya and Nigeria, but in most countries detergents now dominate, exclusively so in most European countries and North America. Yet even in India almost a million tonnes of laundry soap are still manufactured, although even here the amount of detergent now exceeds 1.5 million tonnes a year.

A significant benefit to the environment would come if we all washed our clothes in cold water, something the Japanese have always done. If all washes in Europe were done in cold water then it would save the output of 10 power stations, and in the US it would save energy to the value of around $6 billion per year.

Fabric Softeners

Fabric softeners are cationic surfactants and these are often added to the final rinse of the laundry and dispensed to the rinse water via a special compartment in the washing machine. How can a surfactant make clothes feel so soft? The outside layer of fibres tend to be negatively charged, which is why it is best to use anionic surfactants to wash clothes because these molecules will not end up clinging as residues to fabrics. Cationic surfactants on the other hand will be attracted to the fibres and cling to them. The hydrocarbon tails of these molecules then act to stop the fibres matting together while giving them a feeling that is soft and smooth to the touch because each fibre has in effect been given a layer of light oil one molecule thick. While bedding and garments rinsed with fabric softeners feel softer, it is best not to use fabric softeners when washing towels because the surfactant coating makes it harder for them to absorb water, although without the use of a little fabric softener towels in hard water areas become rather harsh.

Cationic surfactants are also the basis of commonly used antiseptic ointments – see box.

The Hidden Benefits of Cationic Surfactants

Cationic surfactants have two features that enable them to attack and destroy bacteria: their positively charged head and their long hydrocarbon tail. The former means the molecule will be attracted to the cell wall of a bacterium, which is negatively charged, the latter means that it can rupture the cell wall by piercing it. When that happens the cell begins to leak and the bacterium dies. The cationic compounds that behave in this manner are more commonly known as quaternary ammonium salts, or quats for short. The quats most used are benzalkonium chloride and cetrimide, and there are antiseptic ointments which rely on these compounds for their skin-healing properties.

Two-in-one detergents based on a combination of anionic and cationic surfactants are not possible because when anionic surfactants, which do the cleaning, are mixed with cationic surfactants, which do the softening, the molecules are attracted to each other thus rendering them ineffective for both cleaning and softening. However, a non-ionic surfactant and a cationic softener combination do not cancel each other's benefits out, and some two-in-one brands of this kind were on sale in the early 1980s in the US, but lower costs moved the industry back towards anionic surfactants and put paid to such products. More recently, detergent companies have again been selling two-in-one combinations which contain chemicals which don't interact. There are also 2-in-1 products which rely on an entirely different type of fabric softener: the common mineral bentonite, and these have been available for more than 20 years. This is a soft clay consisting of very fine flakes of a magnesium aluminium silicate. When these are included in a detergent they stick to fibres giving them a soft feel in just the same way as the mineral talc of talcum powder imparts softness to skin.

All Washed Up?

The modern dishwash tablet is almost as complex as the modern detergent and there are some ingredients that are common to both products because they are tackling the same kinds of food residues on crockery, cooking utensils, and cutlery, that also need to be washed off clothes, kitchen cloths, and table linen. In addition they may have to face the same problem of hard water. However, there are other considerations to be taken into account because items are washed in a dishwasher many more times than an individual garment is likely to be washed, and over time this can result in metals becoming discoloured, and especially those containing silver, while glassware becomes tarnished with a cloudy iridescence.

Dishwash tablets have undergone a transformation in the past few years, going from a simple powder or liquid gel, to two-in-one tablets, to three-in-one tablets, to four-in-one tablets, and recently to five-in-one tablets, which were launched in 2005. This latest version comes with three sections sealed inside water soluble compartments made of the →polymer poly(vinyl alcohol). The various ingredients in a dishwash product have to be kept separate to prevent them from reacting chemically with one another. Some of the ingredients can coexist without reacting, such as the surfactants and the enzymes, as they do in laundry detergent powders, and some are most useful if they are released near the end of the wash, such as rinse aids.

A key ingredient in a typical dishwash cleaner is the water softener. In some machines there is a separate water softening compartment which is charged with salt and where the calcium and magnesium ions in the incoming water are exchanged for sodium ions which don't form insoluble residues. All-in-one tablets generally contain sodium tripolyphosphate (STPP) to act as the water softener, one of the few uses remaining for this previous stalwart of laundry detergents. Cleaning is easier if the water is slightly alkaline, especially if burnt-on food is to be tackled, and to this end some sodium silicate and sodium carbonate are also added.

The trouble with an alkaline wash is that it can etch away glass. Although glass is not soluble in ordinary water it does become slightly soluble in alkali and over time this will cause the glass to develop an iridescent hue on its surface. While this in no way affects the utility of the glass vessel it looks rather as if the glass has not been rinsed properly but this is not so, and such vessels are perfectly serviceable although they are likely to be discarded. It is possible to counteract this slow but persistent attack on glass by having a zinc salt present in the wash water and this is generally added to the dishwash tablets in the form of soluble zinc acetate or zinc carbonate. Now when the sodium ions on the surface of the glass dissolve into the alkaline wash water they are replace by zinc ions which form an insoluble zinc silicate thereby protecting it from further attack. Of course when an item of glassware is washed hundreds of times then even zinc cannot protect it completely and everyday objects like tumblers will eventually suffer, but for less frequently washed items like flower vases and cut-glass tableware then the zinc will serve them well.

With the water softened and the glassware protected, then the other ingredients can get to work: the surfactants to remove grease, the enzymes to remove protein and starch residues, and bleach to remove stains left by tea, coffee, wine and fruit juices. One of the difficulties in choosing a dishwash surfactant is that it will encounter four different types of material in the dishwasher, namely ceramics, glass, plastics, and metals. The surfactant has to grab hold of the food stains and carry them away, but there is a conflict, because the best cleaning agent for ceramics and glass is not the best for plastics. The surfaces of ceramics and glass are negatively charged so it is best not to use a cationic surfactant which would adhere to them and hinder cleaning. On the other hand the outside of plastic is positively charged, for which the reverse applies, i.e. an anionic surfactant would interfere with stain removal. This is why plastic containers are difficult to wet when they are washed with an anionic surfactant. Clearly a non-ionic surfactant is the one to use. Cleaning metal is generally less of a problem; chemical attack is the threat to these items. To protect any silverware that is being washed then benzotriazole[75] is added and this has

75) Chemical formula $C_6H_5N_3$.

the ability to stick to the surface of silver and provide a temporary protective layer against the action of the bleach molecules.

Dishwasher surfactants must not generate a lot of foam because that can block the pipes within the machine, as you may have discovered when, having run out of dishwasher detergent, you have simply tried using some hand washing-up liquid instead. The surfactants in dishwash products are low-foam non-ionic ones such as fatty alcohol ethoxylates, especially those in which the hydrocarbon chain is 12–14 carbon atoms long. These produce almost no foam.

The bleach in a dishwasher tablet or powder is percarbonate and this needs a bleach activator, but now it is possible to add a metal compound as a catalyst because there is no danger of the contents of the dishwasher suffering in the way that some fabrics can be affected. In some tablets a new self-activating bleaching agent, PAP (short for 6-(phthalimido)peroxyhexanoic acid) is used and this needs no activator or catalyst.

At the end of the wash process comes the rinse cycle and here again a surfactant is needed but not to do more cleaning but to reduce the surface tension of the water so that it 'sheets' off the washed items and doesn't leave droplets which dry out and lead to 'spotting', which is particularly noticeable on glassware.

You may be thinking that, wonderful as the dishwash detergents are in making dishwashing no longer a daily chore, we are still wasting energy and natural resources thereby making the planet pay for our decreased workload. Not so. In fact the modern dishwasher appears to be as environmentally friendly as washing by hand, and the reason is that it uses much less water, the heating of which accounts for a significant part of the cost of this twice daily ritual. On a cradle-to-grave analysis of all the inputs and outputs, be they water, energy, and chemicals, it is the modern dishwasher that wins by a short head. What made hand dishwashing particularly wasteful, at least in the UK, was misinformed publicity about the possible hazards of dishwash liquid – see box.

What a Waste of Water!

The conventional way of washing-up by hand was to leave washed crockery and utensils to drain dry or wipe them with a linen tea towel. Then, in the late 1980s, it was announced that residues of the surfactant LAS (short for linear alkylbenzene sulfonate) on washed items might be the cause of various stomach complaints. This allegation appeared to be based on research findings and was widely reported. People then began to rinse washed items under running hot water to remove any last trace of the offending chemicals, and this practice has continued in many homes to this day despite the fact that hand dishwash detergents no longer contain LAS. Had those reporting their finding to the media been a little more open they would have revealed that their research was based on observations on only six rats, whose lifetimes had been spent drinking water containing a thousand times more washing up liquid than any human would ever be exposed to.

Malodours and Air Fresheners

Things in the air we breathe can affect us emotionally and bad smells can depress us just as much as pleasant smells can make us feel relaxed. Subtle fragrances can turn our thoughts to romance, whereas malodours can make love the last thing on our mind, as we saw in Chapter 3. Our sense of smell can play an important part in making us feel good about our surroundings which is why air fresheners are to be found in most homes. Most are used to neutralise unwanted smells or to replace them with better odours.

There are three ways of removing an unpleasant smell from a room or confined space and they are either to overpower it with one that is much stronger and pleasanter smelling, or disguise it by incorporating it into a blend of molecules that the nose perceives as pleasant, or physically remove it, by which I mean either pluck it out of the air or destroy it chemically. All these methods are used by products which sell themselves as air fresheners and they work in various way. Some work by simple evaporation from a wick, a candle, or a gel, some by escaping from a device that increases the dispersion of the air freshener by warming it or by incorporating a tiny fan, and some by being sprayed from a canister.

There are five categories of domestic smell that humans find most offensive and wish to get rid of.[76] They are toilet odours, tobacco

76) Unpleasant bodily odours were dealt with in Chapter 3.

smoke, stale cooking smells, pet odours, and mould/musty odours. Each type of smell can be replicated in the laboratory because the dominant malodour molecules have been identified and mixes of these can be used to simulate a particular odour for testing. (Sometimes the best way of testing an air freshener is to use the actual material such as cat urine or cigarette smoke.) What are the molecules that constitute malodours? They are categorised by the chemical groups to which they belong. Some contain a →carboxylic acid group, some are nitrogen-containing compounds, and some are molecules with a bivalent →sulfur atom. This last category refers to a sulfur atom that forms only two chemical bonds to other atoms, and especially to hydrogen or carbon, such as in the famously smelly hydrogen sulfide, H_2S, which was often encountered in chemistry laboratories of old and which gave rise to them being referred to as 'stinks' labs. Molecules which incorporate such sulfur atoms are among the worst smelling ones in the world, as we saw in Chapter 3.

We are aware that toilet odours touch a primeval nerve and those from other people are especially repugnant and our natural inclination is to want to remove them. They are made up of a few strong smelling molecules, such as butyric and isovaleric acids, which are present in human faeces, and formed by bacteria. These chemicals are a significant contributor to lavatory smells, although there are many other molecules present as well, including sulfurous ones. The nitrogen-containing chemical skatole, which was mentioned in Chapter 3, is particularly characteristic of human faeces, and this is used in part to simulate toilet odours.

The easiest way to dispel an offensive odour is to hide it under a much stronger, more pleasing smell until it has dissipated. A better way is to incorporate it as a base note in a mixture which then smells pleasant. This is akin to what perfume houses do when devising a new perfume. This has to consist of top notes, which are fresh smelling fragrances like lemon or grass, middle notes that are often those of heady smelling flowers like irises or orchids, and bottom notes which are very animalistic and smell of things like leather and even body fluids like urine or semen. Musk is just such a bottom note and in the raw state this is so offensive that people are normally repelled by it, yet it is added to many perfumes including the most famous. However, without these bottom notes a perfume has no depth, and the skill of the chemists of perfume houses is in combining ingredients so they blend and evaporate together. Someone wearing a badly concocted perfume might well smell very fresh at the start of an evening out, but would smell very off-putting by the time the party was over. This knowledge of how perfumes blend together various notes is also the basis of how some kinds of air freshener work. The bad smell is incorporated and thereby disguised as part of an overall acceptable fragrance.

Ideally an air freshener should remove the offending smell completely and there are some chemicals that can go a long way towards doing this. For example, triethylene glycol (TEG)[77] is particularly good at absorbing molecules and removing them from the air. This was known as long ago as 1966 but TEG has only recently been incorporated into domestic products, although it has been used in industry for many years. Add around 5% of TEG to an air-freshener spray and it will absorb malodours as the droplets fall to the floor. There is even a product of this kind in the US which is used in hospitals and which even has some ability to remove microbes from the air as well. Its use was approved in 1980 by the US Environmental Protection Agency and they recommend that to be effective such sprays must contain more than 5% of TEG. This solvent best dissolves chemicals that have certain molecular features such as bivalent sulfur atoms and aromatic rings,[78] which many pungent molecules possess. TEG is effective against pet, toilet, and smoke smells although it is not particularly effective against musty odours. To disguise the malodours which TEG does not capture it is necessary to add a small amount of fragrance to the air freshener so that we are left with a lingering odour that is reminiscent of a garden of flowers, or a ferny woodland glade, or just the fresh outdoors.

Cyclodextrins make strong smelling odours disappear by wrapping the offensive molecules inside a tiny cavity. Cyclodextrins are soluble in water and used in products like Febreze which is sprayed on to the source of the offending smell, such as a dog basket, sports shoes, or a chair which has been soiled by something obnoxious. The contaminated object is sprayed until its surface is damp, so that when it dries it is covered with a layer of this remarkable chemical, thus ensuring that no smelly molecules escape into the atmosphere. Cyclodextrins

77) Chemical formula $C_6H_{14}O_4$ with the structure $H(OCH_2CH_2)_3OH$.
78) These have flat rings with multiple bonds; skatole has two such rings.

are cyclic polymers of six, seven, or eight glucose units (known as α-, β-, and γ-cyclodextrins respectively) which are formed from starch by the enzyme *Bacillus macerans amylase*. The cyclodextrins have a structure rather like a wrist band and they can capture and grip any small smelly molecule which comes their way. Eventually the sprayed-on layer of cyclodextrin will be lost, but by then the smell may have disappeared or, if not, the smelly object can be sprayed again.

Zinc ricinoleate also acts as a good odour reducer and it does this by binding to the sulfur and nitrogen atoms of malodour molecules thereby making them involatile and so unable to pollute the air. This compound is being introduced into home and personal care products as well as into detergents. Another agent that is effective at neutralising acidic malodours is sodium bicarbonate[79] and this is added to some air fresheners that are meant to be sprayed into the air.

Of course we can do without air fresheners if we keep our surroundings clean and well ventilated and you may think that these products are merely adding an unnecessary burden to the atmosphere. You might even be sensitive to some of their ingredients although these are tested so as not to contain irritants. Compared to the natural release of volatile molecules from plants, the human contribution to the environment is still tiny and, like the natural chemicals, the molecules of air fresheners are easily oxidised and washed out of the atmosphere by rain.

The products discussed in this chapter have made lives easier and pleasanter and it is inconceivable that we would ever return to a world in which they were no longer available. The challenge for the next generation is to find ways of making them all from renewable resources.

79) This is more properly called sodium hydrogen carbonate, and its chemical formula is $NaHCO_3$.

Issue: Could We Be the Last Generation Effortlessly to Enjoy Clean Clothes and Crockery?

When the flow of fossil fuels ceases, could the products discussed in this chapter be made from sustainable resources? And cause no pollution? The answer is yes, and that even goes for the mineral component phosphate. In many parts of the world phosphate is now extracted from sewage and recycled as mineral phosphates so it will be a sustainable resource. Surfactants can be made from plant materials, and indeed some are already made by linking carbohydrates, such as sugar, to plant oils, such as coconut oil. Both these crops are now frowned upon as foods, the first because it leads to tooth decay, the second because it is a saturated oil and so bad for the heart. Both are blamed for the rise in obesity. Washing ourselves and our clothes with them is an alternative use we can approve of.

Several plant-derived surfactants are now available industrially, such as sucrose esters and alkyl glycosides.[80] The latter consist of a chain of carbohydrate units as the water-seeking head with a fatty acid tail of between 10–18 carbons, the length depending on the intended application. They are already used for some purposes, especially where the surfactant comes into direct contact with human skin. How safe these new surfactants are can be judged by the fact that some are deemed safe to eat, an example being sucrose stearate, which consists of an unsaturated 18-carbon chain attached to a sugar molecule and this is used as an emulsifier for foods. Although this seems to fly in the face of the need to eat less of such foods, the amount used as an emulsifier is small.

Sucrose esters can be tailored to specific needs by the number of hydrocarbon chains that are attached to the sucrose and these surfactants find applications in foods because they are colourless, odourless, and tasteless. For example, to produce a low fat spread with a high water content, then three fatty acid groups might be required to get the right surfactant to blend the margarine and water to a smooth consistency. Adding a sucrose ester to chocolate can prevent the unappetising white bloom of cocoa butter crystals which sometimes forms on its surface. In Europe, sucrose esters are given an E-code number, E473, which means they have been passed as safe to be used as food additives in all the countries of the EU.

But what about their use in detergents? Then we will need to produce them on a much larger scale. The alkyl glycosides are already manufactured at an annual rate of more than 100,000 tonnes. The carbohydrate part can be extracted from corn starch, wheat, or potatoes. Alkyl glycosides are mainly used as co-surfactants, in other words they are added to other surfactants, and they appear to work synergistically so that less of either surfactant is needed to achieve the same result. Lauryl diglycoside, which consist of a 12-carbon tail attached to two glucose molecules as the head, finds use in dishwashing liquids and in liquid detergents for hand washing delicate fabrics, and it is especially selected because it is kind to hands.

Within a generation we might well see the millions of tonnes of surfactants that currently come from fossil fuels being replaced entirely by plant-derived natural chemicals. There could even be combinations of carbohydrates and fatty acids that chemists have yet to discover which might be even better than the surfactants we use today, so that we will need much less of them, and if they worked well in cold water then they would also contribute to saving energy.

80) Also known as alkyl polyglycoside or APG

7
Better Looking (II): The Art of the Chemist

A small arrow printed before a word in the main text indicates that there is more information on that topic in the Glossary. In this chapter there are also footnotes with the chemical names of dye molecules.

News from the Future

The Global Times News, 21 March 2025

Stolen Paintings Hoard Discovered

More than 20 famous paintings stolen from museums and art galleries in the 1990s have been discovered in an isolated country house in Ireland belonging to a wealthy recluse. Police were called when the owner, Professor Declan O'Brien, failed to respond to repeated calls from relatives in Canada. They found the body of the recluse in a badly decomposed state surrounded by many of the masterpieces which were also rotting away. A police spokesperson said that they believed the collector had paid criminals to steal the pictures to order, and that he had targeted oil paintings by artists of the sixteenth and seventeenth centuries.

"The paintings are in a terrible state," said the Director of the Dublin Art Gallery, "they are damp and covered in mould and mildew – but we believe they can be restored. Such have been the advances in art conservation this century that I have every confidence that what can be saved will be saved and even parts that are damaged will be reconstructed using modern chemical techniques, making the restored parts indistinguishable from the rest of the painting."

An up-market greetings card manufacturer is believed to have offered the Irish Government €5 million to enable the restoration project to go ahead.

Page 5: Old paintings contain toxic lead pigments and should be kept in sealed vaults say environmental activists.

This final chapter is also called 'Better Looking' because it is about restoring works of art to their former glory, about looking deeper than the mere surface of objects, and about discovering whether a work of art is genuine. With the aid of modern techniques of chemical analysis it is possible to discover exactly what materials the original artist used and how they were applied. It is also possible to uncover earlier

Better Looking, Better Living, Better Loving. John Emsley

ISBN 978-3-527-31863-6

versions of a painting that perhaps the artist was dissatisfied with, to see what parts later painters have retouched or painted over, and occasionally to find that fraudsters have been involved. In some cases the analyst has to remove a tiny sample for testing, but today these are so small as to be almost invisible to the human eye, or they can be taken from the unseen edges of a canvas. Even that may not be necessary if one of the non-destructive methods of chemical analysis is used. In addition to analysing a picture, chemists in art galleries and museums are now able to return a work of art almost to its original state if that is what is required, although some people see the changes that time has wrought as positive contributions.

The importance of chemistry in art restoration came to my attention when I attended the Royal Society of Chemistry in London in 2003 to view a painting, *The Dream of the Virgin,* by the Italian master Simone di Filippo Benvenuti da Bologna (1330–1399). The painting, dating from around 1370, had been discovered in the Station Hotel in Bologna in 1938, and it was given to the Society of Antiquaries of London where it was to languish for the next 60 years. Much of the picture had been over-painted a dark brown, gold leaf had been applied to parts of it, and the whole covered with several layers of varnish. Only when conservators X-rayed the picture did they realise it had a landscape background. Other investigative techniques showed the Virgin's robe was the rare and expensive pigment ultramarine, and they revealed that Benvenuti had used egg yolk to make his colours, which was the standard method of preparing paint in the 1300s. The conservators removed the layers of varnish and over-painting by means of a scalpel and with the help of a specially devised solvent mixture.[81] By the time the picture was ready for public display it had again become a work of great beauty, and an example of what dedicated chemists can achieve in the service of art.

Paintings begin to change from the day they are completed and while this may not be noticeable during the lifetime of the painter and the patron, they would certainly be shocked if they could see the ultimate ravages of time. Colours fade or darken under the influence of heat, light, dirt, and atmospheric gases, varnishes turn dark brown, and cracks may cover the picture from end to end. These cracks are

81) This consisted of acetone, xylene, and N-methyl-pyrrolidone. [Information kindly supplied by Alan Phenix.]

caused either by the slow drying of the oils with which the artist mixed his palette of colours, or by the expansion and shrinkage of the underlying surface as it responds to temperature and humidity changes in the environment. Some of these changes can be dealt with, but not all. (Paintings on paper are also susceptible to degradation of the cellulose of the paper by acids within the paper.) This chapter is mainly about old masters who painted with oils on a background that was on canvas or wood.

There are four aspects of art restoration where chemical knowledge is all important and these are the topics we will look at in this chapter. They are: colour, analysis, restoration, and fraud.

Colour

We know there are supposed to be seven colours in a rainbow – red, orange, yellow, green, blue, indigo, and violet – but most people recognize only six, and I must confess to being colour blind with respect to violet and indigo, seeing them merely as shades of the same colour. The human eye has receptors that recognise three colours, red, green, and yellow, but by a clever trick of the brain we are able to register four primary colours – red, blue, green and yellow. However, we can distinguish thousands of shades and the artists of old had to replicate these, which they did either by blending paints or by painting one layer of colour on top of another. Today there are hundreds of ready-mixed paints of every description from which artists can choose, and these are based on the work of chemists of the 1800s and 1900s who discovered whole new groups of colourful molecules. Artists particularly welcomed the new and brilliant greens. Previous generations avoided green if possible because there were no naturally green minerals that would retain their colour when ground to a fine powder, and they tended to avoid green dyes because these were unreliable.

Sometimes all that is left of a painting are fragments of plaster found during archaeological digs. Even though they may not tell us much about the work of art they came from they can tell us something of the materials available at the time they were executed. Little has survived of the wall paintings on Roman villas in England but sometimes there is enough to reveal that the pigments those ancient artists used included vermilion, red ochre, yellow ochre, green earth, charcoal, soot, red lead, Egyptian blue, and orpiment. In drier parts of the world, such as around the Mediterranean, whole sections of ancient paintings have come to light, as in the palaces of the Minoans in Crete and the tombs of Ancient Egypt, and they show that artists used the same palette of coloured minerals. By the Middle Ages the great masters also had access to 'lakes', which are organic dyes precipitated on to aluminium hydroxide, but these gave colours that were much less permanent.

Pigments and dyes come either from plants, insects, or minerals, and their sources varied over the centuries. Those from plants and insects are organic molecules which generate colour by reacting with light of the visible spectrum which can excite the outermost electrons. The colour we see depends on the light which they absorb, so that if a dye appears red then we know it is absorbing green and blue light. If it is green, which is how the chlorophyll molecules of plants appear to us, then this is because leaves absorb red and blue light. The downside of this behaviour is that these easily excited electrons tend to be reactive chemically, which is why dyes fade over time as their molecules react with oxygen and ultraviolet light. On the other hand a metal pigment will keep its colours for thousands of years because this generally derives from the chemical state of the metal ion which is not so susceptible to change. When the colour component resides in the non-metal part of the pigment things are less permanent and lead white is the most extreme example of such vulnerability, going from purest white to deepest black as a result of its reaction with the gas hydrogen sulfide – more of this later. First let us look at the traditional pigments and dyes of which red and blue have by far the most interesting stories to tell.

Red

Between the years 1400 and 1890 the reds that artists used were extracted from plants, such as madder, or from crushed insects, such as the scale insect *Dactylopius coccus,* or were of mineral origin, such as cinnabar (mercury sulfide, HgS) which could be ground to a powder to produce the pigment vermilion which was much used in antiquity, or red iron oxide (Fe_2O_3) which was used by the Neolithic cave painters, or minium (red lead, Pb_3O_4).

The painters of the Italian city states such as Florence, Sienna, and Venice, preferred to use reds extracted from insects although they may have had little choice because the supply of the most popular red of sappan wood, imported from the East, was cut off when Constantinople fell in 1453. In 1497 the great German artist, Albrecht Dürer painted a portrait of his father[82] and he used madder red. That same year Michelangelo painted *The Virgin and Child with Saint John and Angels*[83] and the bright red dress of Saint Veronica in the painting is coloured by kermes red derived from insects. In 1533 Hans Holbein the Younger painted *The Ambassadors* and he used lac red, also from insects, and this is what Tintoretto (1518–1594) favoured, as demonstrated by the red drapery on which Venus is sitting in his work *The Origin of the Milky Way*. By the next century one dye triumphed over all others and that was cochineal from scale insects. Carmine, as it was called, was used by the Dutch artist Anthony van Dyck's for his *Charity* (1627) and by the Spanish painter Diego Velázquez for his portrait of Archbishop *Fernando de Valdes* painted in 1640. In the 1700s we know carmine was part of Canaletto's palette, as shown by his *Regatta on the Grand Canal* painted in 1740. Meanwhile the English artists Reynolds (1723–1792) and Gainsborough (1727–1788) also preferred the red of carmine.

Madder (*Rubia tinctorium)* is a wild creeper which was used as a dye in India around 5000 years ago. The ancient Egyptians dyed cloth with madder and so did the Romans, as seen from pieces of cloth excavated from Vindolanda at Hadrian's Wall in the North of England. Madder is also native to the Mediterranean, and once its importance as a dye was realised in Medieval Europe it began to be cultivated in Northern France and the Netherlands. Madder produces the intensely red molecules alizarin[84] and purpurin[85] when its roots are crushed. Most madder production went for dyeing fabrics, but some was available for artists in the form of a lake. Lakes were produced by adding alum, which is potassium aluminium sulfate, to a solution of the dye, followed by soda (sodium carbonate) which makes the solution alkaline and precipitates aluminium hydroxide to which the dye molecules

82) The painting, known as The Painter's Father is said to be his work, but it has been disputed.

83) Unless specified otherwise, the paintings mentioned in this chapter belong to the National Gallery London.

84) Chemical name 1,2-dihydroxyanthraquinone, $C_{14}H_8O_4$.

85) Chemical name 1,2,4-trihydroxyanthraquinone, $C_{14}H_8O_5$.

cling. This was then used as a pigment by artists, and the British painter Turner (1775–1851) was still working with madder in the 1800s.

Lac red is extracted from the ant *Kerria laccia,* an insect which colonises the branches of various trees, such as the sacred fig which grows in India and Southeast Asia. The lac is harvested and crushed to obtain the red dye which is still widely used in those regions. After the dye has been extracted, the material which remains can be treated with alkali to yield shellac, formerly widely used as a varnish and lacquer for furniture, clocks, paintings – and 78 rpm records, when it was invariably coloured black.[86] It requires around 50,000 insects to product a pound of shellac. Lac red was being used as early as 1200 BC in India and it was imported to Europe until the 1600s. It was especially popular with the Florentine painters of the 1400s and was used by Michelangelo in several of his works.

Kermes red may be the oldest known dyestuff and it was obtained from the scale insect of that name, *Kermes vermilio* which is to be found in countries of the southern and eastern Mediterranean. It is mentioned in *The Bible*. Kermesic acid[87] is the natural chemical responsible for the colour and it constitutes 1% of the body weight of the insect. Such was its importance in Roman times that it was part of the tribute paid to the occupying army. Even in the Middle Ages it was so highly valued that landlords were reputed to accept it as payment from tenants, and Roman Catholic cardinals wore cloaks dyed with it, such was its expense and their need to demonstrate the prestige of their rank. When cochineal began to arrive from the New World, the demand for kermes red began to decline although is continued to be used as a dye in Venice up to the 1750s. In 1995 it was thought that *Kermes vermilio* had become a rare insect although in that year a traditional dyer in Tunisia was discovered still to be using it.

The story of cochineal is beautifully told by Amy Butler Greenfield in her book *A Perfect Red,* published in 2005. Cochineal was the dye which the Aztecs had discovered, and it is produced by the scale insect *Dactylopius coccus* that feeds only on the prickly pear cactus. The red molecule of cochineal is carminic acid[88] and it constitutes 10% of their

86) They consisted of 25% shellac mixed with cellulose (from cotton), powdered slate and wax.

87) Chemical name 9,10-dihydro-3,5,6,8-tetrahydroxy-1-methyl-9,10-dioxo-2-anthracenecarboxylic acid, $C_{16}H_{10}O_8$.

88) Chemical name 7-α-D-glucopyranosyl-9,10-dihydro-3,5,6,8-tetrahydroxy-1-methyl-9,10-dioxo-2-anthacenecarboxylic acid, $C_{22}H_{20}O_{13}$.

body weight. Not only that, but the colour is particularly intense and it soon superseded kermes red. The Spaniards who colonised Central America began to export cochineal to Europe where people were prepared to pay highly for it. Meanwhile the colonisers guarded the source from which it came and were more than happy to allow people in Europe to believe that it was extracted from seeds. Some merchant adventurers went in search of its source and although they smuggled some of the insects out of Mexico they found it impossible to breed them. Cochineal was used to dye the uniforms of the British Army, the famous redcoats of imperial history.

Somewhat puzzlingly, cochineal has been detected in a 1300s painting by Nardo di Cione called *Saint John the Baptist with Saint John the Evangelist and Saint James*. The explanation is that there was an Old World cochineal beetle that was native to Poland and this too was collected and used and while it too produces the same carminic acid it lacks the intensity of red that the New World cochineal produced. The market for cochineal remained vibrant for almost 350 years until the coal tar dyes devised by German chemists in the late 1800s began to be preferred, not least because they were cheaper and offered a wider range of reds.

The reds of natural dyes mostly fade with time – but not always. Swedish wall hangings discovered in the village of Överhogdal in 1910, and more than a thousand years old, are now regarded as national treasures. They are on display in the Jämtlands läns museum in Östersund. They depict stories of Viking mythology including the tree of life and Odin's eight-legged horse. The principal coloured thread in the tapestries is red and it is almost as bright today as it was when first embroidered sometime between 800 and 1000 AD (according to carbon-14 dating). The red dye has been identified as madder. The explanation of why it has kept its colour so long is thought to be a combination of a cold climate and protection from sunlight.

The reds of mineral pigments do not normally fade with time but they can change and this is true of vermilion which sometimes goes black for reasons not yet understood. Vermilion was not the only red pigment available to artists. There was also minium (red lead oxide, Pb_3O_4) but that has a somewhat orange hue, although it was used in Europe for painting miniatures, and was particularly popular in China. Another manufactured pigment was iodine red (mercury iodide, HgI_2) which was favoured by Victorian water-colour artists who specialised in botanical illustrations.

Blue

Blue was the colour that fascinated painters for centuries: there was so much blue in Nature, and so little of it to be found as pigments that would retain their colour. There was the blue dye indigo, which came from the common woad plant *Isatis tinctoria* but European painters disliked this because the colour faded quickly. Unknown to them, across the Atlantic the painters of the Mayans were producing stunning murals with indigo, as we can see in the ruins at Chichen-Itza, and these should by now have weathered away. So what has preserved them? The answer is that the Mayan artists added palygorskite clay and this has molecular cages which trapped the indigo molecules, thereby protecting them from the bleaching effects of sunlight and oxygen.

Egyptian Blue is the striking colour of the headgear worn by Queen Nefertiti, wife of Pharaoh Akenhaton (reigned 1353–1336 BC), in the famous bust that was discovered in Tel El Armana. The recipe for this pigment had been lost by the time the Romans conquered Egypt in 30 BC, but more recent analysis showed that it was copper silicate ($CuSiO_3$). It was made from sand, lime, and malachite. If clean white sand were stirred into a paste of lime and water and then powdered malachite added the result was Egyptian Blue. Modern samples of this pigment have been compared to that used by the Egyptians and their X-ray analysis patterns are exactly the same.

Ancient artists knew of two other intensely blue pigments and these were made from the minerals azurite and lapis lazuli. Azurite is a variant of basic copper carbonate. Lapis lazuli is more complex and is a silicate rock with variable amounts of aluminium, sodium, calcium, and sulfur and was known as ultramarine. This mineral is found alongside marble and is produced when rock becomes heated under great pressure. It was mined only at Badakshan, Afghanistan,[89] and such was its rarity in Europe that it was worth its weight in gold. It was imported via Venice and many believed it was being secretly manufactured in that city. Such was the beauty and depth of its colour that

89) More lapis lazuli became available in the 1800s when the rock from a mine near Lake Baikal in Siberia was found to contain it.

Medieval painters, or their patrons, were prepared to pay highly for it. Even earlier, the monks of the AD 800s used ultramarine when painting the exquisite artwork of the Book of Kells, now at Trinity College in Dublin, Ireland.

Ultramarine is a brilliant blue and is not affected by light. Producing ultramarine was a lengthy affair beginning with heating the rock and plunging it into cold water, causing it to fracture and making the fragments easier to grind. Then followed a series of lengthy steps involving mixing the powder with oils and extracting the mass with potash to remove impurities. What was left behind were tiny particles of the blue pigment which were then ground up and kneaded with wax to remove impurities and those particles which were so fine they had lost their colour. (The finer the mineral was ground the paler it became.) Its chemical composition was known as long ago as 1806, having been worked out by Charles-Bernard Desormes and Nicolas Clément at the École Polytechnique in Paris, but the exact nature of the material was not solved until 1929 when its crystal structure was revealed by X-ray analysis.

The first successful recipe for making synthetic ultramarine was devised by Jean Baptiste Guimet in 1824 in response to a reward being offered by the Société d'Encouragement pour L'Industrie Nationale. He used a mixture of china clay, sodium carbonate, and a trace of sulfur, with lesser amounts of silica and rosin or pitch. He heated this slowly to 750 °C and then allowed it to cool while keeping the furnace tightly sealed. Depending on the proportion of the ingredients, various shades of blue could be produced, some with a greenish tint, others with a slight reddish hue. Ultramarine was thereafter available at one tenth the price of the natural pigment.

Prussian Blue is probably the most famous blue pigment. It was discovered by accident in 1704 and is made from potassium ferrocyanide and ferric chloride. Heinrich Diesbach, a colour manufacturer of Berlin, had run out of potash (potassium carbonate) with which to make a red lake so he borrowed some from Johann Dippel an alchemist. While this worked fine, something happened to the solution after he had filtered off the red lake: it turned a deep blue colour. Dippel's potash had been made from calcined bones and these contained cyanide from the decomposition of their protein component and the cyanide had reacted to a deep blue compound which we now know as

Prussian Blue.[90] Soon this was being manufactured as a colouring agent and was a great success.

In the 1800s the French Government encouraged their chemists to create new pigments and as a result a new blue, cobalt blue, was announced in 1802, devised by Guyton de Morveau and Louis Thénard. This is cobalt aluminium oxide[91] and is regarded as a 'pure' blue because it does not have either the slight green or the slight indigo tinge of traditional blues. More recently the blue pigment copper phthalocyanine has become popular with artists.

Yellow

Yellow ochre was used by the cave painters of the last Ice Age. The pigment is hydrated iron oxide[92] and in some locales there are massive outcrops of this mineral, such as that near Rousillon, France. The hundreds of drawings discovered in a cave at Vallon-Pont-d'Arc in the Ardèche in 1995 were carbon-dated from soot on the walls which had come from the torches the artists had used and analysis at three independent laboratories confirmed that the paintings were between 30,000 and 33,000 years old. More recent artists, such as those of ancient Egypt, preferred yellow orpiment (arsenic sulfide, As_2S_3),[93] which is another natural mineral, while painters of the Middle Ages tended to use brilliant lead–tin yellow which was made from lead and tin oxides by a process that its producers kept secret .[94] This pigment was a time bomb for any painting that contained it, as we shall see.

There are other synthetic yellow pigments such as cadmium yellow, which is cadmium sulfide (CdS), and its vivid colour was once very popular although it is no longer used because of the cadmium it contains, which is an accumulative poison. Another bright yellow, in this case based on two toxic elements, is chrome yellow which is lead chromate ($PbCrO_4$) and sold under a variety of names such as Cologne yellow and King's yellow. This occurs naturally as a mineral but that used by artists was manufactured. A rarer pigment was lemon yellow, which is strontium chromate ($SrCrO_4$).

Natural yellow plant dyes such as curcumin and berberine are much less intense, but sometimes the use of such colorants could lead

90) It chemical formula is $Fe_4[Fe(CN)_6]$.

91) Its chemical formula is $CoAlO_3$.

92) Its chemical formula is FeO.OH.

93) The other form of arsenic sulfide is realgar (As_4S_4) which is orange.

94) It composition was Pb_2SnO_4.

to a damaging misunderstanding, as in the fate of one of the most valuable books in the world – see box.

Yellow is Not Always a Sign of Ageing

The *Diamond Sutra* is 1200 years old and is particularly valuable because it was the first ever printed book. It is a scroll five metres long and is dated in a way that corresponds to 11 May 868 AD. It was one of a large library of scrolls stored in a cave near Dunhuang, China, which had been sealed sometime before 1025 and was rediscovered by a Taoist priest Wang Yuan-lu who became its self-appointed guardian. He showed the collection to the British explorer Sir Mark Aurel Stein in 1907, who was allowed to remove 7,000 of the scrolls, which he sent to London. This was only a fraction of the library and today there are parts of it to be found in museums around the world including 10,000 items in the National Library of China in Beijing.

When Sir Mark returned for a third visit in 1913 he collected another 600 scrolls although these are now regarded as forgeries, and indeed so well organized had the local forgers become by then that it is now believed that many of the scrolls in collections around the world are not genuine.

The *Diamond Sutra* is genuine but when it arrived at the British Museum in 1909 it was cut up into separate sheets and these were glued to paper. The front part was then treated with bleach in the belief that its yellow colour was due to ageing. The yellow disappeared but thankfully further bleaching of more pages was not carried out. In fact the yellow was a dye that had been specially used to conform to a Buddhist belief that this colour conferred solemnity. The dye was produced from the Amur cork tree (*Phellodendron amurense*). Modern analytical techniques were used in 1995 to analyse the original Chinese paper and to identify the original printing inks. Present day conservators are now restoring the scroll to its original state.

Green

Green posed a problem for painters in the Middle Ages. There was malachite which was obtained from the green mineral basic copper carbonate[95] and while this is bright green and occurs naturally in many parts of the world it becomes much paler when finely ground. Painters of the 1400s obtained this pigment from the friars of San Guisto alle Mure, who also made a synthetic version. The two varieties can be distinguished in paintings because they have a slightly different particle shape. Malachite replaced the mixtures of blue and yellow which earlier painters used, and it in its turn was replaced by verdigris

95) Its chemical formula is $CuCO_3.Cu(OH)_2$.

in the 1500s because this mixed better with oils. Verdigris had not been used by the earlier Florentine painters because it did not mix well with egg. Verdigris is basic copper acetate[96] and it was made by corroding copper metal with vinegar fumes. While it starts as a light green colour it is chemically unstable and can lose the acetate as acetic acid, and then it turns brown. Malachite will also darken due to the formation of copper sulfide (CuS). Greens available to later artists were emerald green, which is copper acetoarsenite, and more recently copper phthalocyanine.

Purple

Purple paint could be made by mixing a crimson lake with black and a little white, and this is what Rubens used. There was a purple dye, Tyrian purple, also known as royal purple, which was extracted from the shellfish *Murex bradaris* found around the Mediterranean[97] but it was far too expensive. In Roman times it was reserved for dyeing the clothes of the Emperors, hence the term 'wearing the purple' which was used to describe those elevated to that exalted position. Lesser mortals who wanted to wear purple had to wait until the second half of the 1800s when the dye mauveine first appeared, and then purple and mauve becoming extremely popular for all kinds of things. Mauveine had been discovered by an 18-year-old laboratory technician William Perkins in 1856 in his home laboratory in East London and his story is ably told in the book *Mauve* by Simon Garfield. There are now modern purple pigments, such as cobalt violet.

Brown

Brown was obtained by mixing the colours at the red–yellow end of the spectrum. There were several earth pigments, such as native and burnt clays, which yielded ochres and the darker umbers and these were the ones mainly used. Brown pigment was also produced by grinding ancient Egyptian mummies to a fine powder, and this was much prized by the great artists. In the 1800s a synthetic version of this could be made by heating strongly a mixture of pine resin, mas-

96) Its chemical formula is $Cu(CH_3CO_2)_2.Cu(OH)_2$.
97) The purple come from the molecules 6-bromoindigo and 6,6′-dibromoindigo.

tic, and beeswax until it became like pitch. When that was ground it too produced fine shades of brown. Later in the 1800s various coal-tar colorants appeared.

Black

Black was usually lamp black, which was the soot produced by burning oils, tar, pitch or resin in a restricted amount of air.

White

Lead white is basic lead carbonate, and it has been used by artists, painters, and decorators for two thousand years because of its remarkable covering power. In Roman times lead white came from the Mediterranean island of Rhodes where it was made by laying thin strips of the metal across bowls of vinegar and leaving them for several months until they became coated with the pigment which was then scraped off and ground to a fine powder. Lead white was formed in two steps, first the acetic acid fumes from the vinegar attacked the lead to form lead acetate, and then carbon dioxide and water vapour of the air reacted with this to form basic lead carbonate.[98] This method of making lead white persisted for hundreds of years, until the Dutch found a better way of boosting production in the 1600s by carrying out the process in a room in which piles of manure were put alongside the vinegar. The room was then sealed for 90 days, by the end of which all the lead would have been transformed into lead white. The decomposing manure provided warmth and lots of carbon dioxide.

For artists – and house painters – nothing came close to matching the brilliance and depth of lead white and while there were alternatives, made from chalk, calcined bones, oyster shells, or even ground-up pearls, they did not compare to lead white. Today this pigment is rarely used and that which is available is restricted to conservators and restorers. We are now aware how toxic this metal is – see page 204 – and lead white has been replaced by titanium dioxide, which delivers the same brilliant whiteness with no risk to health.

98) Its chemical formula is $2PbCO_3.Pb(OH)_2$.

Oils and Varnishes

Preparing colours was a skill an artist learned while an apprentice and it involved grinding up a pigment with an organic binder which was either egg yolk or an oil. (Egg yolk is a mixture of protein and fats.) Artists of the Italian School used mainly egg at the start of the 1400s but by the end of that century they were mainly using 'drying' oils, although some used a mixture of both. Where egg was still used this was for the under-paint layers or for areas such as skin tones or pale blue sky when a colder white was required. The term drying oil implies evaporation as part of the process by which it hardens but this is not what happens and a better term would be 'curing'. Pigments also affect the curing process, sometimes slowing this down, other times speeding it up. The oils most frequently used in paintings of the 1500 and 1600s were linseed, walnut, and poppy oil, although the last of these cures only slowly, but it was used because the quicker curing oils were also prone to yellowing. The reason oil paints harden is due to the double bonds in their fatty acid chains reacting with oxygen and light thereby linking the chains together. The recipes of Theodore Turquet de Mayerne, which he compiled between 1620 and 1646, mentioned poppy oil several times and indeed it is a very pale oil so would have been good to use with pale coloured pigments. It has been identified in the white ermine in Philippe de Champaigne's *Cardinal Richelieu* painted around 1637.

Some artists favoured a particular oil and used it for all their pigments and we know that Leonardo da Vinci used only walnut oil for all the colours in his *The Virgin of the Rocks* painted in 1508. Other artists used two or more oils, and throughout his life Nicolas Poussin used either walnut or linseed oil, sometimes using both, as in his *Landscape in the Roman Campagna with a Man scooping Water* (1638) and *Landscape of a Man washing his Feet at a Fountain* (1648). Turner was less particular about the materials he used when mixing his colours. In addition to the more common oils he is reported to have used various varnishes as well as beeswax and spermaceti wax (from whales).

Old masters were almost always given a final coat of varnish, partly to enhance the depth of colour (known as colour saturation) and partly as a necessary protective layer in the days when buildings were heated with open fires and stoves, which gave off damaging fumes and smoke. Exactly which oil and varnish an artist used can be determined by modern methods of chemical analysis, and such analysis has been an important contribution that chemists have made to art restoration in the past 25 years.

Analysis

Just how sensitive chemical analysis has become was demonstrated in 1999 on a miniature painting of Queen Elizabeth I. She had presented this to Sir Thomas Heneage in 1600 and it became known as the Armada Jewel. It was analysed by Raman spectroscopy using a laser beam to penetrate the glass cover of the jewel, and directed at various parts of the painting. Its scattered rays were recorded and their telltale wavelengths measured. The scattered rays correspond in energy to vibrations within the pigments of the painting, and they revealed, for example, that the pearl necklace in the painting was lead white with orpiment at its edges, the red flowers were vermilion, their leaves were malachite, and the blue background to the painting was azurite plus ultramarine.

A cross section through a painting will reveal a great deal. The lower layer will be found to be either a wood panel or canvas, on top of which is a layer of plaster called the *ground* and this is made from chalk (calcium carbonate, $CaCO_3$) or one of the forms of calcium sulfate ($CaSO_4$) that can be made from the mineral gypsum.[99] The artist would often sketch the subject on to this white ground using something like charcoal, and then he would paint the picture or portrait and this would generally involve several layers of paint. Finally he would apply a coating of a transparent varnish to protect the painting and to give the colours added depth. How can all these various layers be analysed? In 1956 Joyce Plesters described how pin-sized samples, weighing between 5 and 10 mg, could be removed from the edge of a painting, or from around existing damage, then embedded in polyester resin before being cut at right angles so that their layer structure could be studied using a microscope. A trained eye can then identify some pigments merely from their crystal shape. Modern pigments are

99) Plaster on walls that are to be painted is also a form of calcium sulfate, in this case the dihydrate ($CaSO_4.2H_2O$). The paint is applied to wet plaster so that when it dries the paint is firmly bonded to the surface. The result is a *fresco*.

not hand-ground in the traditional pestle and mortar but industrially milled and consequently are much finer than those of the old masters.

Analysis of a painting is essential if there is any doubt about its provenance, or if there is suspicion that it is not the claimed artist's work. More often analysis is designed to discover the materials the artist used and how he applied them, and to uncover any later additions, as well as to assess the extent of deterioration and identify attempts by previous restorers to rectify them. Chemists have at their disposal several analytical tools. Some are by nature destructive, by which is meant that a sample will be removed and so permanently lost from the painting, but many modern ones are non-destructive like the Raman technique mentioned above. Non-destructive methods of analysis are preferred, and while the term 'destructive analysis' implies permanent damage to a painting this only involves extracting a minute piece that would normally be hidden under the frame, or one that is so tiny that its removal will not be noticed. From such miniscule fragments the modern analytical chemist can deduce an enormous amount of data.

Some destructive techniques require a thousandth of a gram while some need less than a millionth of a gram. One particularly useful combination of techniques is gas chromatography (GC) linked to mass spectrometry (MS), which together can separate the components of a complex mixture and identify all of them. In MS, molecules in the gaseous state are ionized and then exposed to an electric field. This accelerates them to a speed that enables a magnetic field to deflect them and the extent to which they are deflected depends on the atomic mass of the particle which thereby reveals its constituent atoms, from which its molecular structure can then be deduced.

Liquid chromatography is used to separate out the chemical components of dyes so that they can then be identified by spectroscopic analysis. In 1980 the German dye chemist Helmut Schweppe published a method using thin-layer chromatography for separating and identifying a wide range of natural dyes. High performance liquid chromatography (HPLC) is now the preferred technique because it can be coupled to an ultraviolet–visible spectrometer which scans the wavelength range of 250 to 750 nanometres and which can analyse each component in the dye as it is separated out. Alternatively it can be coupled to mass spectrometry to identify the components more accurately.

The first non-destructive way of analysing a painting was to X-ray it, and while this does not identify any of the materials used by the artist it does reveal the under-drawing and any changes that have been made to the picture, and maybe shows that one picture has been painted on top of another. It is the lead of lead white which reveals most information because the more this was mixed with a colour the more opaque the painting will be to X-rays. However, X-rays can do more than merely look at the skeleton of a work of art and in recent years X-ray fluorescence (XRF) has become a key method of non-destructive analysis and with equipment that is portable. This sophisticated technique uses X-rays for analysing elements in small samples which it does by identifying them from a pattern of energies that are unique to a particular element. XRF has exposed ingenious forgeries – see box.

No Pope!

A bronze bust of Pope Paul III, who was born Alessandro Farnese in 1468, was bought by the National Gallery of Art in Washington D.C. and attributed to Guglielmo della Porta, a pupil of Michelangelo's, as were similar busts in other museums. There were rumours that all were fakes and conservation scientist Lisha Glinsman was asked to examine them, which she did using X-ray fluorescence spectroscopy. This revealed them to be made of brass (a copper–zinc alloy) and not bronze (a copper–tin alloy). In itself this was not decisive – both alloys were known in the Middle Ages – but it was what she did *not* find that was conclusive. Copper refining in the time of Pope Paul III, who reined from 1534 to 1549, produced a metal which contained several trace impurities and this is characteristic of the metal of that period. These impurities were not present in the bust because the forger had used copper which had been refined by the electrolytic method introduced in the 1800s and which removes all impurities. Glinsman concluded that the Paul III busts were made in the late 1800s or early 1900s.

Infrared rays, while not so penetrating as X-rays, can be used to identify chemical components and with a suitable camera they have even been used to detect the under-drawing of a painting, especially if the artist has used charcoal, a graphite pencil, or black ink, all of which contain carbon. Infrared reflectography can sometimes distinguish between originals and copies and even copies that were known to come from the same artist's studio. For example both the National Gallery in London and the Louvre in Paris have identical paintings by Marinus van Reymerswale called *Two Tax-Gatherers* which depicts two men whom the artist clearly detested, judging by the way he depicted them. Which of the two paintings is the original? The answer was revealed by infrared inspection which showed the London painting to be an exact copy carefully drawn from a tracing of the original, while the Paris painting is the original in which the style of under-drawing is much freer.

Infrared analysis involves exposing a sample to the full range of infrared radiation and noting wavelengths that are absorbed. Chemical bonds within a molecule vibrate with frequencies that correspond to infrared frequencies and observing which they are will often identify a material. Again only a pinprick size sample is required and the resulting spectrum can be compared to a library of stored spectra to identify the substances in the sample. Like infrared analysis, Raman analysis is also based on molecular vibrations and this technique has been used to identify pigments.

Another micro-destructive technique is laser-induced breakdown spectroscopy (LIBS) which is used to analyse the paint layers. Nanosecond laser pulses vaporise a small amount of material from the surface of the painting and the amounts so lost are only 50 billionths of a gram and too small to be seen with the naked eye. The vapour passes between two high-voltage electrodes which excite the atoms and these then emit a pattern of light energy bands which identify the elements. LIBS together with Raman spectroscopy has been used to examine Russian icons which are multilayered.

Conservation

Modern conservation aims at minimal intervention and seeks to prevent a painting deteriorating in the future. It is impossible to view an old painting in the form in which the artist saw it when it was first completed because of the various chemical changes that have occurred during the intervening years, changes that are due to ultraviolet light, heat, humidity, and gases in the atmosphere. In addition to these accidental changes there are those deliberate changes which people in the past have made, either to protect the painting by applying more varnish, or to repair it by touching up parts that have become damaged. Maybe a work of art should be allowed to age naturally, but what is not acceptable is the accumulated dirt of ages and clumsy attempts at restoration by lesser mortals. Given that these additions are not what we want to see, the question arises as to how to remove them, and then to decide how much conservation and restoration should be carried out. Chemist restorers are now in a position to return a painting almost to its original state. While that would allow us to view the work as the artist intended, and many would welcome that, there are those who see the changes of time as an added attraction to a work of art. Both viewpoints have to be considered.

A case in point was the restoration of the ceiling of the Sistine chapel in Rome which was carried out by the Vatican's conservators, and which revealed a brilliance not seen since the days when Michelangelo painted the great work. So vibrant were his colours that some critics were upset by the result, likening it to a Disney cartoon, although others said it was one of the greatest revelations of the modern age. To remove the grime of the centuries, the conservators had cleaned the ceiling using a process normally reserved for cleaning marble. It involved apply a chemical known as AB57, which consists of sodium and ammonium bicarbonate suspended in a cellulose gel, followed by washing with water.

Michelangelo's colours have withstood the test of time but other old masters have suffered irreparable change because they used dyes which have undergone chemical reactions. We can see this effect in Dutch paintings of the 1600s some of which have trees with blue leaves, although when they were originally painted they were green. The reason is that instead of using a green pigment such as malachite they created the green they wanted by mixing a blue pigment, such as azurite, with a yellow lake of plant origin, based on natural yellow dyes such as weld, safflower or buckthorn. Over the centuries this yellow dye has decomposed leaving blue-leafed trees as in the *Hunting Scene* by Pynacker.[100] For the same reason, the dead bird in Greuze's *A Girl with a Dead Canary*, which is in the National Galleries of Scotland, sits as a white canary in a wreath of blue leaves. Greuze has used a yellow lake for both the bird's plumage and to create the green of the leaves.

Some of the red sunsets painted by Van Gogh have likewise faded to grey because he used a synthetic alizarin. We know that his *The Oise at Auvers* (Tate Gallery, London) originally had a pink sky because this is still the colour on the canvas that had been protected from light by the frame of the picture.

100) This can be seen in the Dulwich Picture Gallery, London.

In the days before photography the rich resorted to portraits as their way of immortalising themselves, or at least letting future generations see what they looked like and how important they were. Thousands of such portraits now adorn the walls of great houses and art galleries and while not all are of high artistic merit they do tell a chemical tale. The greatest British portrait painter was Sir Joshua Reynolds but why do so many of those he painted look so pasty-faced? Undoubtedly some of the gentry of Britain in the 1700s were anaemic, exposed as they were to too much lead in their diet and that causes anaemia, or they may have been simply following the fashion for a pale skin which top people valued because it indicated the class to which they belonged. Probably it was neither, but was due to Reynolds' choice of madder for skin tones. When the red alizarin molecule in madder is oxidised it becomes colourless and this is what has happened. Reynolds persisted in using this lake pigment until late in his career when he was finally won over to using vermilion which will resist the ravages of time.

Not all colours fade, some change in the other direction and that is particularly true of the most commonly used pigment, lead white. When paintings were hung in buildings that were heated with coal, then sulfurous fumes from the burning fuel would react with the lead white and convert it to black lead sulfide (PbS). The same effect can be seen on many ancient manuscripts, where faces that were originally painted pale pink are now deep black. It is possible to return PbS to the original basic lead carbonate but that is a complicated process and was beyond the capabilities of restorers in the past. What they could do was convert the lead sulfide to another lead compound that was white, namely lead sulfate and this was achieved with →hydrogen peroxide which oxidises the sulfide to sulfate. This chemical became commercially available in the late 1800s and it was widely used. Sometimes this treatment did not work because the white lead had been oxidised to black lead oxide (PbO_2) by the action of oxygen and light. However, this can be reconverted back to basic lead carbonate by treating it with a mixture of acetic acid and hydrogen peroxide to form lead acetate which then slowly reverts to the original pigment by absorbing carbon dioxide and water from the air.

Lead is also doing something else to old oil paintings. Tiny white pimples can be seen on many of them although these are only really visible through a magnifying glass. These blemishes have been investigated and shown to be lead salts of fatty acids, which form as the original oils break down. A Belgian conservation scientist, Leopold Kockaert, first investigated these defects and thought they were due to protein from the egg of the binder. Even when the artists had used oil, Kockaert assumed that some egg yolk had been mixed with it. He was wrong, as chemist Catherine Higgitt and colleagues at London's National Gallery were to report in 2003. The tiny white spots came from the oils themselves and these had reacted with the lead pigment to form mainly lead salts of the →carboxylic acids stearic and palmitic, together with azelaic acid which is a product of the decomposition of polyunsaturated fatty acids that are present. Azelaic acid does not form when egg tempera ages, because there is no polyunsaturated fatty acid in fresh eggs. The lead salts were identified using infrared microscopy on samples as small as 10 microns in diameter and they were compressed to a thin film between two diamond windows and the infrared beam shone through the sample to reveal its constituents.

The formation of the white spots was not linked to the use of lead white but to other lead pigments, namely red lead and lead-tin yellow which are thought to be much more chemically reactive. So does all this mean that eventually all the oils of oil paintings will slowly decompose? The answer is 'no' and the experts now believe that while some white spotting is inevitable it is a self-limiting process which will eventually cease before any real visual changes take place.

Restoration

Nowadays restorers use modern materials to make up the paints they use and while they may try to match the colours the original artist used, the new ones have to be ones that can easily be removed by a future restorer if it is deemed to be inappropriate or if a better method of restoration becomes possible. Chemist conservators obtain their paints and varnishes from specialist firms like Gamblin Artists Color Company of Portland, Oregon. As well as being easily removed these new paints must be stable, and suitable for the various artistic styles used by the great artists. And they must be materially *different* from the original paints so that future restorers can know exactly what is original and what is not. Gamblin's range of 33 colours are specially designed for conservators and are based on work done by E. René de la Rie when he was based at the National Gallery of Art in Washington DC. He had previously worked at the Metropolitan Museum of Art in New York City, where he specialised in finding a new varnish to replace those that were removed during restoration. His varnishes were badly needed.

Removing layers of varnish from an old painting can reveal remarkable details and wonderful colours, and this is done using chemical solvents of which there are many but few of which are safe to use because of their powerful solubilising nature. Alan Phenix, of the Getty Conservation Institute in California, has studied more than 40 solvents and compiled charts of those properties which are important for removing discoloured varnishes. Not all solvents are suitable but conservation chemists are now in a better position to strip away old varnish without causing damage to the underlying painting. Richard Wolbers of the University of Delaware, Newark, Delaware, has also devised methods of cleaning paintings and his method is based on water as the solvent and uses detergents and enzymes to assist in removing the varnish and its associated dirt.

Having cleaned and restored a painting, what should be applied to protect it and create the depth of colours of the original varnish? The chemical industry had some newer varnishes to offer such as poly(vinyl acetate) and poly(*n*-butylmethacrylate) and some conservators began to use them before they were really aware of the effect they would have on the underlying painting. Some of these transparent polymers did indeed produce the desired effects but they absorbed oxygen from the atmosphere and this caused the polymer chains to become cross-linked, thereafter making them almost impossible to remove. Thankfully, that does not happen with poly(vinyl acetate). René de la Rie came up with two materials that had the necessary properties when applied as a varnish, and these were hydrogenated hydrocarbon resins and urea-aldehyde resins. These stay transparent and glossy, so they delivered the looked-for colour saturation, they don't crack on ageing, and are easily removed by simple hydrocarbon solvents.

Different chemical skills are needed to repair frescoes. In the Florence region of central Italy there are thousands of them, some badly in need of restoring, yet frescoes are among the most difficult types of painting to restore. Their weakness stems from a lack of cohesion between the base material and the plaster. The fresco technique required the artist to apply a smooth paste of lime plaster to a wall and then to paint on it while it was still damp. Over the following week the surface would absorb carbon dioxide from the air to form calcium carbonate and in the process would trap the pigments and dyes of the painting and preserve them forever – or at least for hundreds of years. What threatens frescoes most is attack by the atmospheric pollutant sulfur dioxide which converts the calcium carbonate to calcium sulfate, whose crystals are much larger, so that a fresco will start to lift from the surface as a powdery bloom. Brush this powder away and you lose a lot of the painting. In the 1970s Piero Baglioni, the professor of chemistry at the University of Florence, was asked for his advice on restoring decaying frescoes and with the help of Dino Dini, an art restorer, he came up with an answer: treat the surface with a solution of ammonium carbonate, which reacts with the damaging calcium sulfate to form soluble ammonium sulfate and leave behind regenerated and insoluble calcium carbonate. It was then possible to wash away the ammonium sulfate. This treatment was followed by an application of barium hydroxide solution which reacts with any remaining sulfate to form barium sulfate which was more compatible with the original surface and protected it to some extent.

In 2000 a new technique for preserving frescoes was announced by Baglioni, by then heading a group of surface chemists from six Italian universities, and all of them experts in various aspects of fresco conservation. Their new technique is to follow the ammonium carbonate treatment with a colloidal dispersion of calcium hydroxide in the solvent 1-propanol.[101] This penetrates the fresco better, and over time it again reacts with carbon dioxide from the air to regenerate calcium carbonate. This technique can restore frescoes that earlier conservators would have deemed almost impossible to save. Sadly some frescoes are beyond even this restorative technique.

One of the greatest of all wall paintings is that by Leonardo da Vinci in the monastery of Santa Maria delle Gracie in Milan. This famous masterpiece, *The Last Supper,* was painted in 1495–1497. Since that time it has suffered somewhat, partly as a result of the materials da Vinci used, partly at the hands of earlier repairers and conservators, partly as a result of the environment, and partly as a result of an air raid in 1943 when the refectory of the monastery, on whose wall it had

101) Its chemical formula is $CH_3CH_2CH_2OH$.

been painted, received a direct hit. Almost miraculously the fresco survived. In fact the work is not technically a fresco because da Vinci did not use the wet plaster technique which requires quick and uninterrupted painting. He wanted a different wall surface, so he first sealed the plaster with a layer of paint. This enabled him to work at a gentler pace but this under layer did not allow the plaster to breathe so the damp that seeped into it could not escape. The result was that his masterpiece showed signs of deterioration after only 20 years.

More damage was done in the 1700s when several artists of the day retouched parts that were flaking off. The painting was varnished in 1726 and this preserved a lot of the original from further damage, but the removal of that varnish in the late 1700s was done using a scraper, which did not help matters. Then in the 1800s an attempt was made to remove the painting to another site away from the damp atmosphere in the refectory but that attempt was soon abandoned. At the end of World War II the painting was covered with waterproof felting and boards and left for three years before the refectory could be rebuilt but by then it was dirty, dark, stained, and its surface swollen. In the 1960s and 1970s more scientific restoration was undertaken but advances in chemistry were beginning to turn restoration from guesswork to science. Today, thanks to the sponsorship of the Italian chemical industry, da Vinci's masterpiece has been restored to something like its original glory, with the missing sections skilfully repaired in a way that meets modern restoration standards. Once again its puzzling representation of the Last Supper can be seen by the thousands who flock to it every year and it has achieved yet more fame thanks to Dan Brown's best-selling novel *The Da Vinci Code* which is based in part on the theory that the fresco contains clues to a disturbing secret about Jesus Christ's sex life.

Frauds and Fakes

A painting *Reading in the Forest* was attributed to the French impressionist Eva Gonzalès, and was a hitherto unknown work. (A genuine painting of hers with this title resides in the Rose Art Museum, Brandeis University, at Waltham, Massachusetts.) X-ray fluorescence analysis at Harvard University and Raman microscopy at University College London was carried out on the alternative painting and showed that the pigments were those in use at the time it was alleged to have been painted in 1879. However, the signature proved it to be a fraud even though the painting was not a fake. Gonzalès's son Jean had sold the painting, which was by another impressionist artist, Edmond-Louis Dupain, to a dealer who had then added Gonzalès's name to it in order to increase its value.

Over-painting can be done to make a work more saleable and a famous case was a supposed portrait of Martin Luther by the much admired Protestant artist Hans Holbein the Younger. That is how the painting was attributed in the 1797 guidebook to the contents of Stowe House, the country seat of the Marquess of Buckingham, where the painting then resided. It is now correctly known as the *Portrait of Alexander Mornauer,* who was the town clerk of Landshut in Bavaria, and it has been dated to around 1470. What puzzled conservators at London's National Gallery, who examined the painting in 1991 using XRF, was that the pigment providing the blue background was Prussian blue. Clearly there had been over-painting at a later date. Below the blue background was a layer of varnish and below that another background, brown in colour. The oils used to apply the two backgrounds were linseed oil for the original layer, which was consistent with the materials in use when the portrait was done, and poppy seed oil for the Prussian Blue.

The painting now resides in the National Gallery London, where a decision was taken to remove the top layer. This revealed that it had not been applied to repair damage to the lower layer because there was no damage. What appeared to have happened was that, for reasons unknown, several inches had been cut off the top of the painting, but this resulted in an unbalanced composition. The hat that the sitter was wearing was now too tall, so its size was reduced by over painting the top half of it so as to match the background. More importantly the writing on a letter held by the sitter could now be read and this said: 'To the honorable and wise Alexander Mornauer, town clerk of Landshut, my good patron'. The sitter was also wearing a ring on his left thumb which shows a moor's head, a pun on the name Mornauer, and he was indeed town clerk of that small Bavarian town for 24 years from 1464 to 1488. Sometime in the 1700s a dealer had changed enough of the picture to make it saleable as a rare portrait of Martin Luther and so it had remained for 200 years or more.

Such is the celebrity of some artists that their work is only affordable by the very rich. This demand attracts other artists who hope to make money by forging paintings, and such is their skill that they can fool even the experts. The people they cannot fool are the chemists and there is now a growing number of these scientists working in the art world, generally in the back rooms of national galleries and forensic laboratories, who are expert in analysing works of art, and occasionally their skills can be used to prove that a work of art is not what it purports to be. In earlier times the lack of analytical techniques made it easier for forgers to succeed and some did spectacularly well.

In 1937 an art dealer Han van Meegeren claimed to have discovered an old painting in a forgotten store room in a castle in Holland and thought it might be by the famous Dutch painter Johannes Vermeer (1632–1675). Art experts hailed it as an early work by the master and a museum in Rotterdam purchased *Christ and His Disciples at Emmaus* for around $250,000. During World War II Holland was overrun by the Nazis and at the end of that war the collection of paintings amassed by Herman Goering was examined and another early Vermeer came to light. It too had been purchased from van Meegeren, who was arrested and charged with treason for selling national assets to the enemy. He then confessed that it was a forgery and that several other Vermeers which had come to light in the war years and been sold by him were also fakes. The authorities were not convinced, maintaining that van Meegeren was merely saying this to escape justice.

During his trial, and to prove what he said was true, van Meegeren demonstrated his painting skills in the courtroom and explained how he had managed to make them look old. He told how he had taken unimportant paintings of the 1600s and removed the artwork, and then painted his Vermeers using exactly the pigments and oils that the master would have used. Finally he placed the finished work in an oven and baked it to harden the pigments and cause an authentic looking pattern of cracks. The charge of treason against van Meegeren was dropped and replaced by one of forgery, of which he was found guilty, but he died in 1947 before he could begin his sentence.

The fake Vermeers were eventually dated to the 1930s and 40s by Bernard Keisch at the Brookhaven National Laboratories in 1968. He analysed the radioactivity in the paintings and showed that the lead white that van Meegeren had used was relatively new. Lead contains a little radioactive uranium-238 which disintegrates through a series of other radioactive isotopes among which is radium-226. Ancient white lead contained minute traces of these isotopes which can be identified by the radioactivity they emit. When modern lead is mined and refined, the radium-226 is removed and so it was for van Meegeren's forgeries.

Would-be forgers of Egyptian papyri were caught out by chemists when they tried to sell six such documents, one of which was purported to be as old as 1200 BC, the rest being of a later date but all supposed to be before 100 BC. The papyri were so delicate that they were mounted under glass and the vendors were not keen for them to be removed from their protective frames. Nor was that necessary, because they could be examined by their Raman spectra which can be recorded from samples held under glass. This identified various pigments such as iron oxide, chalk, and ultramarine which of course the ancient illustrators could have used. What was a little less likely was that the Egyptian artists of 3,000 years ago had independently discovered Prussian Blue, which was also present, along with two phthalocyanine dyes, one blue and one yellow, which were first made in Scotland in the 1930s.

A well-known art forger of the early 1900s was Icilio Federico Joni who specialized in producing paintings of Italian artists of the 1400s and he created what would have been described as masterpieces had they been painted 500 years earlier. Later developments in non-destructive chemical analysis techniques were able to expose his works as frauds by showing that he had used pigments such as cadmium yellow, chrome yellow, cobalt blue, and viridian which were not available in the 1400s. Joni had developed a technique to produce the crazed pattern of cracks of old paintings and this he did by applying a layer of shellac resin, mixed with a little pine resin, to the finished work before he applied a final varnish coating. As the shellac resin hardened it contracted and this produced the cracks like those in medieval paintings. The shellac also gave the game away because this was not available in Europe in the 1400s.

Chemists are now respected members of the art restoration scene, in some cases turning paintings that look good into ones that look even better, in other cases they are helping to expose fraudsters, not only of the present but of the past. Their identification of the pigments used by painters might also explain the behaviour of the great artists, who often suffered from undiagnosed mental conditions. They may have been affected by the materials they were using, as the following reveals.

Issue: Were Great Artists Affected by the Pigments They Used?

The answer is very likely, and pigments are a possible explanation of the way artists were prone to abnormal moodiness and poor health. As we have seen, many pigments were compounds of the toxic elements, mercury, lead, and arsenic. As long ago as 1713, the physician Bernardino Ramazzini speculated that both Correggio and Raphael had been victims of lead poisoning.

The Spanish painter Goya has also been thought to have been affected, as was Van Gogh who was particularly fond of sucking the paint from his brushes, and this would certainly have contained lead. Van Gogh's erratic behaviour and mental state are consistent with the effects of lead poisoning but this metal had its most profound effect on Goya. He was born in 1746 in the Aragonese village of Fuendetotdos, Spain, and trained as an artist. In 1775 he became employed as a designer at the Royal Tapestry Factory and worked there for 17 years, during which time he used excessive amounts of lead white for the backgrounds of the tapestry designs and which he insisted on grinding for himself (to save money). In the 1780s he painted portraits and was a conventional and much sought after artist. Then in 1792 he became seriously ill with all the symptoms of chronic lead poisoning, namely severe colic, constipation, pale complexion, tremors, deafness, temporary blindness, and mental disturbances such as paranoia and fits. He was bedridden for two months. Thankfully he recovered but he was left deaf and his painting changed to a darker style for which he is now best known and exemplified by his most famous work *The 3rd of May 1808: the Execution of the Defenders of Madrid*, which can be seen in the Museo delo Prado in Madrid.

Nor was lead white the only danger to which artists were exposed. Other pigments were also made of lead such as chrome yellow, lead-tin yellow, and red lead, all of which were employed well into the 1900s. Auguste Renoir (1841–1919) was a heavy smoker of hand-rolled cigarettes and through this habit may well have been affected by the pigments he was using, especially ones containing mercury – so wrote Lisbet Pedersen and Henrik Permin, of the University Hospital of Copenhagen, in *The Lancet* in 1988. It would certainly explain why his later works do not stand comparison with his earlier paintings.

Did the eccentricities of the great artists stem from their being victims of low-level poisoning? Lead and mercury are well known to affect the brain. Given the skills of today's forensic chemists it might one day be possible to reveal more about the great painters themselves provided we can find at least a strand of their hair or some part of their remains. Or maybe it would be better not to know. Finding lead and mercury in a great artist's remains might remove some of the mystique of his vision of the world and may be taking chemistry in the service of art just a step too far.

Glossary

Words in *italics* means there is also more information under these headings.

ABS is short for poly(acrylonitrile-butadiene-styrene) and this is a *copolymer*, so called because it is made from a mixture of basic monomer units, each of which bring some desirable property to the final product. ABS is widely used to make children's toys, car facias, and fingernail extensions.

Acrylonitrile, butadiene and styrene can be polymerised individually. Acrylonitrile yields polyacrylonitrile which makes soft fibres for clothing (sold as Orlon and Courtelle), while butadiene forms polybutadiene or synthetic rubber, and styrene as polystyrene is known for its excellent insulation properties.

Aluminium chloride comes in two forms: anhydrous aluminium chloride (chemical formula Al_2Cl_6) which is dangerous to handle on account of its reaction with water when it generates a lot of heat, and aluminium chloride hexahydrate (chemical formula $Al_2Cl_6.6H_2O$) which is safe, although it behaves as a weak acid when dissolved in water. There is also basic aluminium chloride which has some of the chlorides replaced by hydroxides (OH) and there are two variants of this which are commonly used: $Al_2(OH)_5Cl_2$ and $Al_2(OH)_4Cl_2$. They are often referred to as aluminium chlorohydrate and they are neither dangerous nor acidic.

Better Looking, Better Living, Better Loving. John Emsley

ISBN 978-3-527-31863-6

Anaesthetics mentioned in this book are:

Anaesthetic	Chemical formula	Boiling point (°C)*
Chloroform	$CHCl_3$	61
Desflurane	$CF_3CHFOCHF_2$	25
Diethyl ether ('ether')	$CH_3CH_2OCH_2CH_3$	35
Enflurane	CHF_2OCF_2CHFCl	56.5
Halothane	$CF_3CHClBr$	50
Isoflurane	$CF_3CHClOCHF_2$	58.5
Nitrous oxide	N_2O	(gas) -88
Servoflurane	$(CF_3)_2CHOCH_2CH_2F$	58.5

* The lower the boiling point the higher the vapour pressure at room temperature.

Betaines – see *quaternary ammonium compounds.*

Carbohydrates were misnamed due to a misunderstanding by early chemists. The general chemical formula for carbohydrate is $C_6H_{12}O_6$ which approximates to $6C + 6H_2O$ and it is this apparent combination of the element carbon and water that led two Frenchmen, Joseph Gay-Lussac (1778–1850) and Louis Thénard (1777–1875), to give them the name carbo-hydrate, but this in no way relates to their chemical behaviour. The alternative name for carbohydrates is saccharides, which comes from *saccharum* the Latin word for sweet. 'Glyco' is also used as a prefix to indicate a carbohydrate grouping and this is derived from the Greek word *glukus,* meaning sweet. So, for example, the term glycoprotein refers to a carbohydrate linked to a protein, and glycolipid refers to carbohydrate linked to a fat (lipid) molecule.

The most abundant carbohydrate in Nature is glucose and this is a monosaccharide of chemical formula $C_6H_{12}O_6$. The other common monosaccharides are fructose and galactose. When two monosaccharides join together then we have a disaccharide ($C_{12}H_{24}O_{12}$) such as sucrose, which is glucose + fructose and is ordinary sugar. Other disaccharides are lactose which is glucose + galactose, and maltose which is glucose + glucose. Oligosaccharides consist of chains of three or more monosaccharides, and polysaccharides consist of many hundreds of monosaccharides linked together as a polymer chain, of which starches and cellulose are common examples.

Carboxylic acids are characterised by having a carbon atom bonded to two oxygen atoms, one of which holds the acidic hydrogen. The chemical formula for such a group is CO_2H. The simplest carboxylic acid is formic acid (HCO_2H), which is its traditional name, although chemists prefer to call it methanoic acid. The following table lists some of the more common carboxylic acids in alphabetical order of their more commonly used names:

Common name	Chemical name	Chemical formula	Natural source
Acetic acid	Ethanoic acid	CH_3CO_2H	Vinegar
Arachidonic acid	Eicosatetraenoic acid	$C_{19}H_{31}CO_2H$*	Essential fatty acid
Azelaic acid	Nonanedioic acid	$HO_2C(CH_2)_7CO_2H$	Rancid oils
Benzoic acid	Benzoic acid	$C_6H_5CO_2H$	Fruit berries
Butyric acid	Butanoic acid	$CH_3CH_2CH_2CO_2H$	Sweat
Caproic acid	Hexanoic acid	$CH_3(CH_2)_4CO_2H$	Goat's cheese
DHA	Docosahexaenoic acid	$C_{21}H_{31}CO_2H$*	Fish oils (omega-3)
Formic acid	Methanoic acid	HCO_2H	Ants
Isovaleric acid	3-Methylbutanoic acid	$(CH_3)_2CHCH_2CO_2H$	Hops, tobacco
Lactic acid	2-Hydroxypropanoic acid	$CH_3CH(OH)CO_2H$	Milk
Oleic acid	9-Octadecanoic acid	$CH_3(CH_2)_7CH{=}CH(CH_2)_7CO_2H$	Olive oil
Palmitic acid	Hexadecanoic acid	$CH_3(CH_2)_{14}CO_2H$	Fatty acids from oils and fats
Parabens	4-Hydoxybenzoic acid	$HOC_6H_4CO_2H$	Strawberries and grapes
Salicylic acid	2-Hydroxybenzoic acid	$C_6H_4(OH)CO_2H$	Wintergreen leaves
Stearic acid	Octadecanoic acid	$CH_3(CH_2)_{16}CO_2H$	Fatty acids from oils and fats
Sweat acid	3-Methyl-2-hexenoic acid	$CH_3CH_2CH_2C(CH_3){=}CHCO_2H$	Armpits
Trifluoroacetic acid	Trifluoroethanoic acid	CF_3CO_2H	Synthetic chemical

* Arachidonic acid has 4 double bonds along the hydrocarbon chain; DHA has 6.

Chirality is the ability of a molecule to exist in two forms which differ only in that one is the mirror image of the other. When it comes to designing a new drug it is generally found that only one of the mirror images is the active form. That is the one which will interact with the enzyme whose malfunction is causing the disease. (The other form might be responsible for unwanted side effects.) Nowadays the aim is only to give the active form and there are companies which specialise in making medicines that have the requisite chirality.

Chitin consists of chains of cyclic carbohydrate molecules linked together and it is basically the same as cellulose except that in place of a hydroxy group (OH) on the rings there is an acetylamino group ($NHCOCH_3$). Like cellulose, chitin is insoluble in water. Removal of the acetyl part to leave the amino group produces a modified chitin which can be used to remove the cloudiness of beers and fruit juices.

It can also be turned into thin sheets like cellophane (which is made from cellulose) and these are used to make slow-release capsules as well as packaging, which is especially useful because it has anti-microbial properties and so helps preserve the food it is protecting.

Colorants. In 1983 the US Congress passed the Federal Food, Drug, and Cosmetic Act popularly known as the FD&C. This permits only seven food colorants to be used in the US, and these are as follows, with their European E number:

FD&C code	EU code	Chemical name	Shade
Blue no. 1	E133	Brilliant blue FCF	Bright blue
Blue no. 2	E132	Indigo carmine	Royal blue
Green no. 3	E143	Fast green FCF	Sea green
Red no. 3	E127	Erythosine	Cherry red
Red no. 40	n.a.	Allura red AC	Orange-red
Yellow no. 5	E102	Tartrazine*	Lemon yellow
Yellow no. 6	E110	Sunset yellow	Orange

* This has been accused of causing hyperactivity in young children, but there is no scientific proof that it does.

Colorants for other purposes, such as hair dyes, are not so strictly regulated. The ones used in semi-permanent dyes are 2-nitro-1,4-diaminobenzene, (orange/red), 2-amino-4-nitrophenol (yellow), 1,4-diaminoanthraquinone (violet), and 1,4,5,8-tetra-aminoanthaquinone (blue), and a combination of these will produce the desired shade.

Contact lens materials – see *gas permeable polymer*.

Copolymer – see *polymers*.

Double-blind tests – these are conducted under conditions where neither the recipient nor the person giving out the drug knows whether the medication is the test material or a placebo.

Drug trials. Before a medicine is licensed it has to undergo rigorous testing. A new substance is first tested on laboratory animals such as rodents to ensure it is not toxic and to see if it cures the disease, and then on larger animals and finally on primates. If it passes these tests then it is tested on humans and this is done in four phases. Phase I involves a small group of healthy volunteers to check that it has no detrimental side effects. Phase II involves larger groups of individuals (100–200) who are suffering from the disease to check that it really is a beneficial treatment. If these results are positive then it moves on to Phase III trials when it is given under medical supervision to a larger number of patients, as many as 3,000. Finally the drug receives a license from the authorities and it can then be made available to doctors to prescribe to patients, this is now Phase IV. At this stage it is still monitored because rarer side effects might still come to light.

Electrical energy is measured in watts (W) per hour. A low energy light bulb uses around 10 watts, an electric heater around 2000 watts, which is the same as saying 2 kilowatts (kW). A million watts is a megawatt (MW) and a billion is a gigawatt (GW).

Epidemiology is the study of human populations with a view to finding factors which affect the pattern of occurrence of diseases and thereby uncover a cause or an influential factor which may not be obvious. Epidemiological studies compare two large groups of people, with ideally at least a thousand people in each group. One group consists of those who are affected by the disease in question, while the other group is free of that disease. In all other respects the two groups must be identical in terms of race, sex, age distribution, social class, location, diet, drinking and smoking habits, etc. Carried out by professionals, epidemiology has the status of a serious investigative science. Unfortunately a lot of epidemiology is carried out by amateurs and their findings are barely more scientific than anecdotal evidence or urban myth. Not surprisingly, those conducting such studies generally arrive at results which confirm what they were hoping to find. Often hidden factors have not been taken into account, thereby leading to wrong conclusions. These hidden factors are referred to as confounding variables, not least of which may be an assumption that those who have been questioned have infallible memories and always speak the truth.

Ether – see *anaesthetics.*

Ethoxy is the name given to the grouping CH_2CH_2O.

Fluorocarbons are carbon compounds in which hydrogens have been replaced by fluorine atoms. For example the methyl group is CH_3 and this would become the trifluoromethyl group CF_3. Replacing hydrogens with fluorine has fundamental effects on the properties of a molecule such as making it non-flammable.

Gamma-hydroxy-butyrate (GHB) is also known as oxybate and is derived from gamma-hydroxy-butyric acid which is a neurotransmitter and has the chemical formula $CH_2(OH)CH_2CH_2CO_2H$. It is available as its sodium salt, sodium gamma-hydroxybutyrate. GHB acts to relax the central nervous system thereby making people feel less inhibited. The liver disposes of GHB by rapidly converting it to carbon dioxide and water. GHB is used to treat the sleep disorder narcolepsy and is given in doses of up to 5 grams.

Gas permeable polymers are based on the *polymer* poly(methacrylic acid) to whose chains are attached some silicone groups, such as tris(trimethylsiloxysilane) [chemical formula $CH_2CH_2CH_2Si(OSi(CH_3)_3)_3$]. These make the polymer able to absorb oxygen from the air.

Gigawatt, GW is a measure of *electrical power* and is a billion (10^9) watts.

HEMA is the polymer formed from 2-hydroxyethyl methacrylate, which has the chemical formula $CH_2{=}C(CH_3)(CO_2CH_2CH_2OH)$.

Hydrogen peroxide has the chemical formula H_2O_2 and consists of two linked oxygen atoms each with a hydrogen atom attached. It has a powerful oxidising action and is used as a bleaching agent. See also *urea*.

IR, infrared light – see *light*.

Isotretinoin is a variant of tretinoin, which is better known as vitamin A acid, and has the chemical name 3,7-dimethyl-9-(2,6,6-trimethyl-1-cyclohexen-1-yl)-2,4,6,8-nonatetraenoic acid. Its structure contains a chain of 9 carbons with double bonds between alternative carbon pairs along the chain. Tretinoin has *trans* double bonds which means they give the chain a systematic zig-zag arrangement whereas isotretinoin has *cis* double bonds which produce a more irregular structure.

Keratin is a natural single-strand *polymer* consisting of chains of amino acids, with a high proportion of the two sulfur-containing amino acids cysteine and methionine. Keratin is the polymer of hair and nails. The sulfur atoms account for keratin's ability to bond to arsenic, mercury, and lead. Hair analysis has played an important role in forensic evidence in cases of poisoning, and it has been possible to diagnose the extent to which famous people in history were affected by these toxic elements provided a sample of their hair has been preserved.

Latex is a natural *polymer* made from the sap of rubber trees which contains the monomer isoprene and this polymerises to give latex. The yield of sap can be stimulated by synthetic plant hormones, and it is a sustainable resource, production of which is around 8 million tonnes a year. Isoprene itself is a volatile liquid, boiling point 34 °C, with the chemical formula C_5H_8, and the molecule consists of a chain of four carbon atoms with a double bond at each end and a methyl (CH_3) group attached to one of the inner carbons [CH_2=C(CH_3)–CH=CH_2]. Isoprene, like all such double-bonded materials will polymerise under the influence of catalysts, and in the natural state it is the oxygen molecules of the air that trigger this process. Latex consists of long hydrocarbon chains that still contain a double bond which results in chains with kinks along their length. The chains can be straightened by stretching but then they relax to their original arrangement when released. Natural latex rubber is very soft but it can be made harder by cross-linking the polymer chains via sulfur-containing molecules, a process known as vulcanization.

Lauryl is the old name for a chain of twelve carbon atoms; the modern name is dodecyl (do = 2, decyl = 10). The commonly used surfactant sodium lauryl sulfate is made from lauryl alcohol which comes from coconut oil or palm kernel oil.

Light is that part of the Sun's electromagnetic spectrum which reaches the surface of the Earth, and is visible. There are also rays that are invisible. This spectrum of electromagnetic radiation is divided up as follows:

Wavelength (nm)*	Colour	Effects
200–285	Ultraviolet-C	Invisible and damaging to living cells
280–302	Ultraviolet-B	Invisible and causes skin to darken
320–390	Ultraviolet-A	Invisible and beneficial (produces vitamin D in skin)
390–445	Violet	Visible
445–500	Blue	Visible
500–575	Green	Visible
575–585	Yellow	Visible
585–620	Orange	Visible
620–740	Red	Visible
740 and above	Infrared	Invisible but sensed as warmth

* These units are nanometres, of which there are a million in a millimetre.

Nonoxynol-9 consists of a hydrocarbon chain (C_9H_{19}) attached to a benzene ring to which is also attached a chain made up of nine linked *ethoxy* groups (OCH_2CH_2) terminating in a hydroxyl (OH). The formula is $CH_3(CH_2)_8C_6H_4(OCH_2CH_2)OH$. It is sold under a variety of names such as C-Film, Encare, Intercept, etc.

Oligosaccharides – see *carbohydrates.*

Parabens is the common name for a groups of chemicals based on *para*-hydroxybenzoic acid chemical formula $HOC_6H_4CO_2H$. (Chemists call it 4-hydroxybenzoic acid.) This acid is not used as such but it converted to an ester by substituting an organic group in place of the acid hydrogen of the CO_2H group, so there are the benzyl, isobutyl, butyl, n-propyl, ethyl and methyl esters. These are referred to as benzylparabens, isobutylparabens, etc. and they are added to a product to kill any bacteria that might contaminate it. *Para*-hydroxybenzoic acid occurs naturally in strawberries and grapes.

pH is the term used to describe acidity and alkalinity, and it is really a measure of the concentration of the molecular ion species H_3O^+ (often just written as H^+) which is the active component formed by acids when they dissolve in water. The concentration of this ion can vary so widely that a special scale, known as pH, is needed to measure it. The pH is a logarithmic scale in which this concentration is expressed in terms of negative powers of ten. For example, in neutral water the concentration of H_3O^+ is very low, only 10^{-7} ions per litre, and its pH is defined as 7. The pH scale is an inverse scale as far as acids are concerned, in other words the lower the pH, the stronger the acid. The range of normal acidities is from 1 (the strongest) to 7 (neutral water) and being logarithmic this means that an acid of pH 1 has a million times more H_3O^+ than does neutral water of pH 7.

The pH scale can also be extended to cover alkaline conditions from 7 to 14 when OH^- predominates and the concentration of H_3O^+ falls by a further factor of a million. The full span of pH from 1 to 14 represents a numerical difference of a *million million*. Nevertheless, we can come across materials in our everyday life which encompass the full range:

The Acids and Alkalis of Everyday Experience

pH		Typical substances	Active chemical
0	acidic	chemical reagent	concentrated sulfuric acid
1		stomach acid	dilute hydrochloric acid
2		lemon juice	citric acid
3		vinegar	acetic acid
4		tomato juice	ascorbic acid (vitamin C)
5		beer, rain water	carbonic acid (H_2CO_3)
6		milk	lactic acid
7	neutral	blood	
8	alkaline	sea water	calcium carbonate
9		bicarbonate	sodium hydrogen carbonate
10		milk of magnesia	magnesium hydroxide
11		household ammonia	ammonia (NH_3)
12		garden lime	calcium hydroxide
13		drain cleaner	sodium hydroxide
14		caustic soda	concentrated sodium hydroxide

PMMA is the polymer formed from the monomer starting material methyl methacrylic acid which has the chemical formula $CH_2 = C(CH_3)(CO_2CH_3)$.

Polymers consist of long chains of atoms with the same repeating unit and the basic chemical from which they are made is referred to as the monomer; Greek *mono* = one, *poly* = many. There are biologically derived polymers and synthetic ones.

Typical biopolymers are hair, which is a long chain molecule, *keratin*, formed by the joining together of amino acids, or paper, which consists of cellulose and is made of chains of glucose molecules, or rubber, which is a natural polymer of the hydrocarbon molecule isoprene. While biopolymers like these are still much used, the modern world relies mainly on synthetic polymers.

The most common molecules from which polymers are made are ethylene (chemical formula CH_2=CH_2) and its derivatives. The key to its polymerisation is the double bond, which opens to form bonds to other ethylenes and the end result is a chain of CH_2 groups, –CH_2–CH_2–CH_2–CH_2–CH_2– which can be millions of carbons long. This is polyethylene. Another common polymer is poly(vinyl chloride) which is made from vinyl chloride (chemical formula CH_2=CHCl) and is better known as PVC. See also *PPMA* and *HEMA*. The following table lists various common polymers headed by those based on ethylene and its derivatives:

Polymer	Acronym	Applications
Polyethylene	PE	Depends on the density:
Low density polyethylene	LDPE	Plastic film, bags, coating for paper
High density polyethylene	HDPE	Moulded products, containers, crates
Poly(methyl methacrylic acid)	PMMA	Illuminated signs, hospital incubators, car lights, etc.
Polypropylene	PP	Film, carpets, thermal clothing, bottles
Polystyrene	PS	Packaging, toys, cutlery, drinking glasses
Poly(vinyl chloride)	PVC	Window frames, flooring, tubing, pipes
Poly(ethylene terephthalate)	PET	Bottles, food trays, duvet filling
Butadiene rubber	BR	Tyres, tennis balls
Styrene butadiene rubber	SBR	Tyres, footwear, moulded goods, bitumen
Isoprene rubber	IR	Tyres, footwear, moulded goods, paints
Polyurethane	PU	Foam padding, surfaces, elastomers
Polyketone	PK	Barrier, extrusion, moulded goods
Epoxy resins	ER	Surface coatings, adhesives, composites
Poly(vinyl pyrrolidone)	PVP	Personal care products

PVP is short for the *polymer* poly(vinyl pyrrolidone).[102] When PVP is left in the open air it absorbs around 15% of its weight as water from the air, such is its ability to cling to water molecules. It is also used in various personal care products like shampoos, toothpastes, and is the adhesive which has replace animal-based glues on envelopes and postage stamps.

Quaternary ammonium compounds ('Quats') are based on the ammonium ion, NH_4^+, which is an ammonia (NH_3) molecule with an added hydrogen ion H^+. The hydrogens of NH_4^+ can be replaced by organic groups and these are known as quaternary ammonium compounds. They need a balancing negative charge such as a chloride ion. In theory there can be millions of possible compounds, although only a few are commercially important. When three of the four groups are methyl groups and the fourth is an acid then the compound is electrically neutral and is called betaine, i.e. $(CH_3)_3N^+CH_2CO_2^-$. Again there are many possible betaines depending on the groups on the nitrogen, but only a few are important. Quaternary ammonium compounds are used as surfactants, fabric conditioners, and antiseptics.

Silicones are compounds consisting of chains or rings of alternate silicon and oxygen atoms, and the most commonly used ones are those with each silicon having two methyl groups attached: and the basic chemical unit is $-Si(CH_3)_2O-$. Silicones are extensively used as waterproof sealants.

Sulfur compounds give rise to notoriously smelly molecules when the sulfurs have either hydrogen or carbon atoms bonded to them. The following are the ones mentioned in this book:

Chemical name	Chemical formula	Natural source
Hydrogen sulfide	H_2S	Decomposing proteins
Methanethiol	CH_3SH	Bad breath
Dimethyl sulfide	CH_3SCH_3	Truffles
Dimethyl disulfide	CH_3SSCH_3	Titam Arum lily
3-Methyl-3-sulfanyl-hexan-1-ol	$CH_3CH_2CH_2C(CH_3)(SH)CH_2CH_2OH$	Armpit bacteria

102) Sometimes spelt pyrrolidinone.

T-cells are a type of white blood cell and they play a key role in the immune system. They are formed in the bone marrow and stored in the spleen and lymph nodes. T-cells can live for up to 4 years.

Triclosan is a chlorinated derivative of phenol and its chemical name is 5-chloro-2-(2,4-dichlorophenoxy)phenol and its chemical formula is $C_{12}H_7Cl_3O_2$. Phenol is a benzene ring with a hydroxy group attached, chemical formula C_6H_5OH, and triclosan has two such rings in the molecule.

Ultraviolet, UV – see *light*.

Units of weight used by chemists must span the range from molecules to minerals. The ones that are most commonly used are as follows:

Unit	Symbol	Scale	Weight	Approximate size
Nanogram	ng	10^{-9}	Billionth of a gram	Invisible
Microgram	µg	10^{-6}	Millionth of a gram	Speck of dust
Milligram	mg	10^{-3}	Thousandth of a gram	Grain of sand
Gram	g	1	1 gram	A peanut
Kilogram	kg	10^3	A thousand grams	1 litre bottle of water
Tonne	t	10^6	A million grams	A cubic metre tank of water

Urea has the chemical formula $(NH_2)_2CO$ and is also known by its older name carbamide. Urea is manufactured industrially from ammonia (NH_3) and carbon dioxide (CO_2) and is used mainly as a fertilizer. It is also added to animal feeds to increase their nitrogen content, and is used to make resins, plastics, and pharmaceuticals. Urea peroxide is simply a combination of urea and *hydrogen peroxide* and as such is a useable source of this latter chemical in the form of a stable white crystalline solid.

Index

III/V semiconductors 147
2-hydroxyethyl methacrylate 211
3-methyl-2-hexenoic acid 77
3-methyl-3-sulfanyl-hexan-1-ol 77
3-methylbutanoic acid 69
4-ethyloctanoic acid 76
4-hydroxybenzoic acid 35
6-(phthalimido) peroxyhexanoic acid 171

a

AB57 cleaner 195
ABS – see poly(acrylonitrile-butadiene-styrene)
ACE anaesthetic 61
Acetic acid 69, 97, 117
Acetone 32, 178
Acetonitrile 33
Acitretin 45
Acne 38, 40–42
Acrylamide in foods 123–4
Actimel 115
Activ glass 140
Acuvue Advance 24
Adrenal gland 40
Air fresheners 172–5
Albumin 47
Alcohol 60
Alefacept 45
Algae 109
Alivisatos, Paul 133
Alizarin 181, 195, 196
Alkyl glycosides 74, 176
Allergic reactions 35, 86
Allicin 113
Alliin 113
Alopexil 19
Alpha rays 126
Alphaderm 43
Alum 181
Aluminium chloride 78, 205
Aluminium compounds 75, 78, 79
Aluminium hydroxide 80, 181
Alzheimer's disease 79–80
Amanita muscaria 101
Amase Institute 103
American Cancer Society 14
American Rheumatism Association 48
Americium 126
Amines 69
Aminopeptidase N 122
Ammonia 11, 68
Ammonium bicarbonate 195
Ammonium carbonate 199
Amnesia 91
Amphoteric surfactants 160
Amylase enzymes 157
Amylose 99
Anaesthetics 60–64, 206
Anaphylactic shock 11
Androstenol 77
Androstenone 70, 71, 76–7
Anionic surfactants 160
Anthracene 39
Anti-aluminium campaign 79–80
Antibodies 58, 104
Anti-corrosion inhibitors 158
Anti-malarial vaccine 59
Antiperspirants 75, 78–81
Anti-redeposition agents 158, 163–4
Antiseptic ointments 168
Apocrine gland 39, 76, 78
Arachidonic acid 117
Arid Cream 78
Armada Jewel 191

Better Looking, Better Living, Better Loving. John Emsley

ISBN 978-3-527-31863-6

Armpit 71, 83
Arnold, William 82
Arsenic sulfide 186
Arthritis 46–54
Artificial nails 31–34
Artificial silk, wash damage 166
Artists affected by pigments 204
Aspirin 46, 51
Association of Anaesthetists 60
Astigmatism 20
Atkin's diet 96
Atopic eczema 43
Auranofin 47, 49
Autoimmune disease 46
Azathioprine 44
Aztec's red dye 182
Azurite 184, 195

b

Bacillus macerans amylase 175
Backpack solar panels 132
Bacteria 168
Bacterial carbohydrate 59–60
Baglioni, Piero 199
Baldness 16
Bandgap 148
Basic copper carbonate 184, 187, 188
Basic lead carbonate 189
Bavachee plant 38
Bavarian Solarpark 135
Becquerel, Alexandre-Edmond 129
Beer 101
Benecol 111
Bentonite mineral 169
Benvenuti 178
Benzoic acid 113
Benzotriazole 170–1
Benzoyl peroxide 33, 41
Bergström, Sune 50
Berliner, David 72
Betaines 206, 216
Betamethasone esters 43
Betnovate 43
Bifidobacteria 114, 116
Black pigments 189
Bleach activators 158, 164–6
Bleachable stains 165
Blood types 98
Blue pigments 184–6
Blue-green algae pollution 162
Blu-ray technology 148, 150
Body odour 104
Book of Kells 185
Boron 130
Bouillabaisse 118
BP Solar 128, 135
Bread 98, 101
Breast cancer 58, 80, 81
British Museum 187
Brookhaven National Laboratories 202–3
Brown pigments 188–9
Brown, Nathan 79
Burton, Fiona 91
Butane 75
Butyl stearate 31
Butyric acid 97, 117, 173

c

C_{60} semiconductor 133
Cadmium 126
Cadmium selenide 133
Cadmium sulfide 132
Cadmium telluride 131
Cadmium yellow 203
Calcipotriol 40, 45
Calcitriol 40
Calcium 56, 120
Calcium binders 162
Calcium carbonate 26, 191, 196
Calcium phosphate 26, 27
Calcium sulfate 191
California Public Utilities Commission 128
Campylobacter 116
Canaletto 181
Cancer 11, 58
Candida 34, 43
Cannabis 91
Caproic acid 69
Capsaicin 119–120, 121
Capsicum frutescens 119
Carbamide – see urea
Carbamide peroxide 28
Carbohydrate drugs 54–60
Carbohydrates 95–101
Carbon dioxide 61, 97
Carboxylic acids 69, 76, 173, 207
Carboxymethyl cellulose 27
Carmine 181

Carminic acid 182–3
Cartilage 48
Casanova 84
Castor bean 56
Cationic surfactants 160, 167–9
Catullus 76
Cave paintings 83, 180, 186
Cell membranes 56
Cellobiose 97
Cell-to-cell communication 98
Cellulose 54, 58, 163, 195
Ceramic laundry disks 155
CFCs – see chlorofluorocarbons
Chalk 191
Champaigne, Philippe de 190
Chemical vapour deposition 138
Chemotherapy 16, 121
Chen, George 144
Chernobyl 109
Chewing gum 29
Chichen-Itza paintings 184
Chicory root 116
Chilli 119
Chirality 68, 208
Chitin 54, 208
Chloracne 40
Chloral hydrate 90
Chlorofluorocarbons 75
Chloroform 61, 62, 82, 206
Chlorophyll 126
Chloroquine 49
Cholera 59
Cholesterol 41, 76, 111
Choline 77
Chrome yellow 186
Chrysotherapy 47
Cinnabar 180
Cinnamon 121
Citric acid 162
Clark, Anthony 77
Clearasil 41
Clément, Nicolas 185
Clobetasol propionate 43
Clostridium 114, 116
Cloves 121
Coal tar products 39, 44, 183
Cobalt aluminium oxide 186
Cobalt blue 186, 203
Cocaine 91
Cocamides 160
Cochineal 181
Coconut oil 176, 212
Codeine 48
Codling moth 70
Coindet, Jean-Francois 108
Coindet's tincture 108
Cold menthol receptor type 1 120
Collagen 39
Cologne yellow 186
Colon 58, 113, 114
Colorants, US approved 208
Colour in paintings 179–180
Colour saturation 190
Columbus eggs 117
Comstock Law 85
Concentrated photovoltaic power 134
Condoms 83–88
Conducting polymer P3HT 133
Conjugated linoleic acid 113
Conservation of paintings 194–7
Constipation 55
Contact dermatitis 43
Contact lenses 20–25
Copper acetoarsenite 188
Copper silicate 184
Copper-indium-diselenide 127, 131
Correggio 204
Corynebacterium 76, 77
Cosmetic regulations 75
Cosmetics ingredients 35
Co-surfactants 176
COX 50–52
Cranberries 113
Crawford, John 21
Crest 30
Cretinism 110
Crocin 119
Crowe, John 103
Cyclodextrin 59, 174–5
Cyclokines 53
Cyclopropane 62
Cyclosporine 44
Cysteine 211
Cytokines 45

d

Danone 115
Date-rape drugs 90–92
Davy, Humphry 60
De Navarre, M.G. 79

Defelice, Stephen 111
Dentine 27, 30
Deodorants 81–83
Dermis 39
Dermovate 43
Desflurane 63, 206
Desormes, Charles-Bernard 185
Detergent builders 158, 161–3
Diachylon plasters 89
Diamond Sutra 186
Diaphragms 88
Diarrhoea 115
Dickinson, Michelle 28
Diesbach, Heinrich 185
Diethyl ether 206
Diiodomethane 109
Dimethyl disulfide 70
Dimethyl sulfide 68, 69
Dimethyl trisulfide 70
Dini, Dino 199
Dinwoodie, Thomas 135
Diode junction 127
Dioxane 73
Dioxins 73, 82
Dippel, Johann 185
Disaccharides 206
Disease-modifying anti-rheumatic drug 49–50
Dishwasher detergents 169–172
Dithranol 44
Diuretic 28
DMARD – see disease-modifying anti-rheumatic drug
Docosahexaenoic acid 117
Double-blind tests 167, 209
Dr Power's French Preventative 85
Drug trials 40, 51, 209
Dunhuang cave discovery 187
Dupain, Edmond-Louis 201
Dürer 181
Durex 85
Duron 87
DVDs 148
Dye transfer inhibitors 158, 164
Dye-sensitised solar cells 133

e

E.coli O157:H7 56, 121
Ebbett, Virginia 82
Eccrine gland 39, 76
Ecstasy 91
Eczema 32, 38, 42–44
Egg yolk binder 178
Eggs with DHA 117
Egyptian blue 184
Egyptian mummy brown 188
Egyptian papyri forgeries 202
Electric energy units 209
Elias, Jack 55
Elisha, prophet 16
Emerald green 188
Emperor moth 70
Emulsifiers 161
Enbrel 53
Endorphins 121
Enflurane 206
Enterobacteriaceae 114
Enzymes 35, 54, 104, 157, 170
EPA 123, 174
Epidemiology 209–210
Epidermis 39
Erythromycin 41
Etanercept 45, 53
Ether 61, 62, 206
Ethinyloestradiol 89
Ethoxy group 160, 210
Ethyl acetate 32
Ethyl chloride 62
Ethyl cyanoacrylate 33
Etretinate 45
Eucalyptol 120
Eugenol 71, 121
Eumelanin 10
Everdry 78

f

Fabric conditioners 160, 167–9
Faeces 114
Fake paintings 200–1
Falcarinol 113
Fallopius, Gabrielle 83–4
False nails 31–34
False teeth 25
Farthing, Tom 144
Fatty acids 161
Fatty alcohol ethoxylates 171
FDA 19, 52, 75, 81, 112, 123
Febreze 174
Federal Food, Drug, and Cosmetic Act 208

Feldman, Marc 45
Fernández-Santana, Violeta 58
Fiber K 86
Fibre 54, 95
Finasteride 17–18
Findeker, Jobst 118
Fingernails 31–34
Firmenich 77
Fish odour syndrome 77–8
Flosulide 51
Flunitrazepam 91
Fluorescing agents 158, 166–7
Fluoride 27, 30
Fluoroapatite 27, 30
Fluorocarbons 161, 210
Fluoroether 24
Fly agaric 101
Foam regulators 158, 161
Foam suppressants 161
Folic acid 119
Food calories 1
Forensic Science Service London 91
Forestier, Jacques 46
Fragrances 158
Fray, Derek 144
Frescoes 198–9
Fresnel lenses 134
Frick, Adolph 21
Fructo-oligosaccharides 116
Fructose 58, 95, 97, 116
Fuller, Roy 115
Functional foods 110–114

g

GaInP – see gallium indium phosphide
Gainsborough 181
Galacto-oligosaccharides 116
Galactose 116
Gallium 126, 127
Gallium arsenide 127, 131, 134, 147, 149
Gallium indium phosphide 127
Gallium nitride 147–151
Gallium phosphide 147
Gamma-hydroxybutyric acid 91–2, 210
GaN – see gallium nitride
Garfield, Simon 188
Garlic 113, 121
Gas chromatography 192
Gas-permeable lens 23, 210
Gatorade 101
Gay-Lussac, Joseph 206
Gelatin 100
Gentiobiose 97
Germanium 127, 148
Getty Conservation Institute 198
GHB – see Gamma-hydroxybutyric acid
Gibson, Glenn 116
Gigawatt 210
Gillette 81
Ginger 120
Girassol project 135
Glass 137–140
Glass surface iridescence 170
Glaxolide 71
Glinsman, Lisha 193
Globo-H 58
Glucosamine 53–54
Glucose 57, 58, 59, 95, 97, 98, 116
Glycemic index 96
Glycerol 27, 32, 157, 161
Glycogen 95, 97, 100
Glycolipid 206
Glycoprotein 206
Goa powder 39
Goat pheromone 76–7
Goat tallow 156
Goering, Herman 202
Goitre 106, 108, 109, 110
Gold cyanide 47
Gold sodium thiomalate 46
Gold therapy 46
Golden hamster pheromone 70
Gonzales, Eva 200–1
Goodyear, Charles 85
Gorett, T. 79
Goya 204
Grätzel cell 133
Grätzel, Michael 133
Great Lakes pollution 162
Green pigments 187–8
Greenfield, Amy Butler 182
Greenpeace 127
Gregor, William 142
Growth factors 39
Guggenheim Museum 145
Guglielmo della Porta 193
Guimet, Jean Baptiste 185
Gypsum 191

h
Habaneros peppers 119–20
Haemophilius influenza type b 58
Hahnemman, Samuel 65
Hair 10, 15–20, 39
Hair dyes 10–15, 208
Halothane 62–3, 206
Hard water 163
Havenaar, Robert 115
Hayashibara, Takanobu 103
Heart attacks 52
Helicobacter pylori 121
HEMA – see hydroxyethyl methyl methacrylate
Henna 15 14
Heparin 55
Herpes simplex 43
Hexachlorophene 81–2
Hib – see haemophilius influenzae type b 58
Hickman, Henry Hill 61
Higgitt, Catherine 197
High performance liquid chromatography 192
Hippocrates 17, 26, 45
HIV 88
Hofmann, Thomas 118
Holbein the Younger 181, 201
Homeopathy 65
Homes, Oliver Wendell 60
Honey 101
Honeydew 103
Hooper, Lee 113
House dust mites 43
HPLC – see high performance liquid chromatography
Human surfactants 159
Humectant 27
Humira 53
Hunter, M.A. 142
Hydrocortisone 43, 44
Hydrogel 22
Hydrogen peroxide 23, 27–28, 56, 164, 196, 211
Hydrogen sulfide 68, 173, 180
Hydrogenated hydrocarbon resin 198
Hydroxyapatite 27, 30, 31
Hydroxyethyl methyl methacrylate 22–23, 211

i
Ibuprofen 48
Ilmenite 143
In't Veld, Jos Huis 115
Indica rice 103
Indigo 15, 184
Indium 126, 127
Indium arsenide 147
Indium phosphide 131, 147
Indium tin oxide 132
Indium-gallium nitride 150
Infant rotavirus diarrhea 116
Inflammatory bowel disease 115
Infrared analysis 193–4
Institute of Earth Sciences 146
Institute of Solar Energy 134
International Council for Iodine Deficiency Disorders 106
Iodide 105–110
Iodine 105–110
Iodine antiseptic 105, 108
Iodine deficiency 106
Iodine during pregnancy 107
Iodine red 183
Iodine-131 109
Iodised salt 105, 106
Iodomethane 109
Ionic liquid crystals 133
Iron 117
Iron titanium oxide 143
Isoamyl acetate 71
Isoflurane 63, 206
Isolauri, Erika 116
Isomaltose 97
Isoprene 211–2
Isopropanol 32
Isotretinoin 41, 42
Isovaleric acid 173

j
Jacobson's organ 71
Jessen, George 22
Joni, Icilio Federico 203

k
K-19 dye 134
Keisch, Bernard 202
Keratin 16, 19, 211
Kermes red 181
Kermesic acid 182

Ketamine 91
Ketones 97
Killian, Frederck 85
Kim, Peter 52
King's yellow 186
Klebsiella pneumoniae 34
Klemm, Dieter 59
Kockaert, Leopold 197
Kroll process 144
Kwon, Ho Jeong 122

l

Lac red 182
Lactobacillus 114
Lactoferrin 117
Lactose intolerance 115
Lanolin 39, 81
Lapis lazuli 184
Laser-induced breakdown spectroscopy 194
Latex 85, 211–2
Laundry aids 154–169
Lauryl digylcoside detergent 176, 212
Lauryl trimethyl ammonium chloride 160
Lawrence Berkeley National Laboratory 151
Lead acetate 12, 89, 189, 196
Lead chromate 186
Lead poisoning 89
Lead sulfide 196
Lead white 189, 204
Lecanora 102
Lecithin 161
Lecoq de Boisbaudran 150
LEDs – see light emitting diodes
Leeks 116
Leeuwenhoek, Anthony van 29
Lemon yellow pigment 186
Leonardo da Vinci 21, 190, 199
Leukaemia 13
Ley, Steve 57
Lichen 102
Liebig 90
Light bulb filaments 142
Light emitting diodes 147, 148
Light rays 129, 212
Lim, Drahoslav 22
Limonene 68
Linear alkyl benzene sulfonate 172
Liothyronine 107
Lipase enzymes 157
Liquid chromatography 192
Liquid paraffin 43
Listeria 116
Long chain fatty acids 157
Long, Crawford Williamson 61
Loniten 19
Lonolox 19
Low fat spread 176
Lucite 21
Lucozade Sport 101
Luctulose 116
Luther portrait fake 201
Lycra 86–7
Lysine 118

m

MacMillan, David 57
Madder 180, 181, 183, 196
Magnesium aluminium silicate 169
Magnetic resonance imaging 64
Maini, Ravinder 45
Malachite 187, 195
Malaria 122
Malodours 172–3
Maltose 97
Manganese catalyst 165–6
Manna 101–105
Mannose 98
Margarine 111
Marine, David 108
Mass spectrometry 192
Mauveine 188
Mayonnaise 97
McNeil, Christopher 82
Meegeren, van Han 202
Melanin 10, 39
Melanoma 58
Melanoocytes 39
Mendeleyev, Dimitri 150
Menecrates 89
Menstruation 89
Menthol 120
Menthone 89
Mercury iodide 183–4
Mercury sulfide pigment 180
Metchnikoff, Elie 114
Methacrylate polymers 32
Methotrexate 49

Methoxsalen 38, 45
Methyl cyanoacrylate 32
Methyl mercaptan 69
Metropolitan Museum of Art 198
Micelles 160
Michelangelo 181, 182, 193, 195
Mickey Finn 90
Micrococcus luteus 76
Mifepristone 89–90
Milk 108, 116–7
Mineral oil 43
Minium 180, 183
Minoxidil 18–19
Mirick, Dana 80
Misoprostol 49
Mitsuoka, Tomotari 115
Mobile phones 126
Modified cellulose 85
Mohn scale of hardness 146
Monoamine oxidase 78
Monosaccharides 206
Mornauer, Alexander 201
Mortierella alpine 117
Morton, William 60, 61
Morveau, Guyton de 186
Mottram, Donald 124
Moura photovoltaic array 135
MRI – see magnetic resonance imaging 64
Mucus 50, 117
Muhler, Joseph 30
Muller, F.A. Dr 21
Multijunction solar cells 127
Mum deodorant 81
Murcott, Toby 65
Murex bradaris 188
Muscone 70
Mushrooms 101
Musk 70, 173–4
Myocrisin 46

n

Nail varnish 31
Nakamura, Shuji 148
Narcolepsy 210
Nardo di Cione 183
Narinder Dev 11
National Gallery London 197, 201
National Gallery of Art in Washington DC 193, 198
National Institute for Clinical Excellence 52
Natural ingredient debate 35
Naturalamb condoms 85
Neutraceutical 111
Nickel allergy 43
Nicotinamide 41
Nitinol 146
Nitric acid 27, 108
Nitrocellulose 31, 32
Nitrogen-based malodours 173
Nitrosamines 73, 88
Nitrous oxide 60, 61, 63, 206
n-layer 130
N-methyl-pyrrolidone 178
Nobile, Arthur 50
Nonanolybenzene sulfonic acid 165
Non-biological detergents 158
Non-ionic surfactants 74, 160, 171
Nonoxynol-9 87, 212
Non-steroid anti-inflammatory drugs (NSAID) 49, 50, 51
Nörenberg, Ralf 30
Norgestrel 89
Northrup, Alan 58
Nose 71
Novartis 20, 23, 53
NO_x pollution 141
Nujol 43

o

Oats 156
Obesity 96
Oestrogen 89
Oils used by artists 190
Oligosaccharides 116, 206
Omega-3 fatty acids 113, 117
Omega-6 fatty acids 117
Onions 113
Opium 60
Organic semiconductors 132
Organochlorine compounds 82–3
Osteoarthritis 47
Owen, Thomas 103

p

Paella 118
Painting, *The Last Supper* 199–200
Paintings 191–4
Palm kernel oil 212

Palmitic acid 157, 197
Palygorskite clay 184
Pancreatin 23
Papain 23
Paprika 120
Parabens 35, 81, 213
Paracelsus 60
Paracetamol 48
Paraphenylenediamine 11–13
Pectin 99
Pedersen, Lisbet 204
PEG – see polyethyleneglycol
Penicillamine 49
Pentoxifylline 55
Pentyl acetate 32
Pepper spray 121
Peppers 119
Peracetic acid 165
Percarbonate bleach 171
Perkins, William 188
Permin, Henrik 204
Peroxide bleaches 158, 164–6
Persil 157, 158, 164, 166
Perspex 21
Pfaus, James 72
pH 29, 95, 117, 161, 213–4
Phaeomelanin 10
Phenix, Alan 198
Phennyroyal 89
Phenylene diamine 88
Pheromones 70, 71
Phosphorus 130
Photosynthesis 126
Photovoltaic power 126–137
Pigments 180
Pimecrolimus 44
Plaque 25
Plasmodium falciparum 122
Plasticizer 31
p-layer 130
Plesters, Joyce 191
Plexiglas 21
Pliny 156
Pocari Sweat 101
Poly(acrylonitrile-butadiene-styrene
32, 205
Poly(*n*-butylmethacryate) 198
Poly(vinyl acetate) 198
Poly(vinyl alcohol) 169
Poly(vinyl pyrrolidone) 24, 164, 215
Polyethyleneglycol 27
Polymers 214–5
Polymethyl methacrylate 21–22
Polyphenols 27
Polysaccharides 206
Polyunsaturated fatty acids 117
Pomegranate 113
Porcelain teeth 26, 29, 32
Potassium aluminium sulfate 181
Potassium carbonate 156, 185
Potassium hydroxide 32
Potassium iodate 107
Potassium iodide 107, 108
Poussin, Nicolas 190
Power plant condensers 143
PowerGlaz 137–8
PPD – see paraphenylenediamine
PPMA – see poly(methyl methacrylate)
Praline 58
Prebiotic foods 114–118
Prednisolone 50
Premature babies 159
Preti, George 71
Prexidil 19
Prexige 53
Prickly pear cactus 182
Priestley, Joseph 60
Prince Charles 142
Probiotic foods 114–118
Progesterone 90
Progestogen 89
Promethium 125
Pronyl-lysine 118
Propane 75
Propecia 17,
Propionibacterium acnes 40, 41
Propionic acid 97
Propylene glycol 80
Proscar 18
Prostaglandins 48, 49, 50
Prostate 17–18, 58
Protease enzymes 157
Prussian blue 185, 201, 203
Pseudomonas aeruginosa 34
Psoriasis 38, 44–46
PT-141 72
Pulegone 89
Purple pigments 188
Purpurin 181
PVP – see poly(vinyl pyrrolidone)

Pynacker 195
Pyrogallol 15

q
Quantum well 150
Quaternary ammonium compounds 168, 215–6
Quats – see quaternary ammonium compounds
Quimi-Hib 58

r
Radar 126
Radioactivity of lead pigments 202–3
Raisio 111
Raman microscopy 200–1
Raman spectroscopy 191, 192, 202
Ramazzini, Bernardino 204
Rand, John 26
Raphael 204
Rare metals scarcity 152
Red dyes and pigments 180–4
Regaine 17–19
Reid, Ted 24
Remicade 53
René de la Rie 197
Renoir 204
Resistant starch 98
Restoration of paintings 197–200
Resurrection plants 101
Reymerswale, Marinus van 193–4
Reynolds 181, 196
Rheumatoid arthritis 47
Rice 109
Ricin 56
Ridaura 47
Right Guard 81
Rogaine 17
Rohypnol 91–2
Romag 137
Roser, Bruce 104
Rosin 32
Royal jelly 95
Rubens 188
Rule, Krista 82
Ruthenium 126, 133
Rutile 143

s
Saccharide chemistry 206
Saccharin 27
Saffron 118
Safranal 119
Salicylic acid 41
Saliva 29
Salmonella 116
Salt 30, 74, 107
Samuelson, Bengt 50
San Guisto alle Mure 187
Sanderson, Kevin 140
Santa Maria delle Gracie monastery 199
Sapa 89
Saponin 155–6
Sappan wood red dye 181
Sapphire 150
Scale insect dye 180, 182
Scheduled drugs 92
Schiller, Friedrich 59
Schueller, Eugène 10
Schweppe, Helmut 192
Scientific Committee of Cosmetic Products (EU) 14
Scott-Ham, Michael 91
Scoville scale of hotness 119–20
Scoville, Wilbur 119
Seaweed ash iodide 108, 109
Sebaceous gland 39, 41
Sebum 39, 40, 41, 73, 157
Selenium 24
Semen 71, 173
Sensodyne 30
Servoflurane 63, 206
Sexual desire 72
Sexually transmitted disease 83
Shampoo 74
Shaving cream 73
Shellac 182
Shirota, Minoru 114
Shitake mushrooms 101
Shivers, Joseph 86
Silicon bandgap 148
Silicon dioxide 26, 162
Silicon 127, 129, 132
Silicones 23, 87, 161, 216
Silverware, protection of 170–1
Simmons, D 50
Simpson, John 61
Sistine chapel restoration 195
Sitostanol ester 111
Skatole 68, 173

Smart windows 126
Smell 68
Smith, Ian 78
Smoke detectors 126
Soap 73, 156, 157, 161, 167
Soapwort plant 156
Sodium alginate 27
Sodium alkyl benzene sulfonate 73
Sodium benzoate 27
Sodium bicarbonate 175, 195
Sodium carbonate 156, 170, 181
Sodium carboxymethyl cellulose 163
Sodium lactate 74
Sodium laureth sulfate 73, 74
Sodium lauryl sulfate 25, 26
Sodium monofluorophosphate 30
Sodium perborate 157
Sodium percarbonate 164
Sodium silicate 157, 162, 169
Sodium tripolyphosphate 162, 169–70
Soflens 23
Solar panels 126–137
Solar water heaters 136
Solaronix 134
Solvent extraction 124
Sorbitol 27
Spandex 86
Sperm 56, 89
Squire, Balmanno 39
Staphylococcus 43, 59, 76, 77
Starch 95
Stearic acid 157, 197
Stein, Mark Aurel 187
Steroids 40, 76
Stirling engines 137
Stomach 114
Streptococcus 43
Stroke 52
Strontium 30
Strontium carbonate 186
Styrene-ethylene-butylene polymer 87
Suckling, Charles 62
Sucrose stearate 176
Sugar 25, 54, 96, 116
Sulfasalazine 49
Sulfuric acid 146
Superglue 32–33
Superoxide free radicals 140
Surfactants, and uses 73, 158, 159–161
Sweat glands 76
Swedish National Food Administration 123–4
Sweets 103
Syphilis 84

t

Tacalcitol 40
Tacrolimus 44
Tall oil 111
Tazarotene 45
T-cells 43, 44, 216
Tea 30
Tea tree oil 74
Technetium 125
Teeth 25–31
TEG – see triethylene glycol
Testosterone 16–17
Tetracycline antibiotics 27, 41
Thénard, Louis 186, 206
Thermus thermophilus 100
Thin film photovoltaics 132
Thromboxane 51
Thyroid 107
Thyroxine 107
Tin dioxide on glass 138, 139
Tintoretto 181
Titan arum lily 68
Titanium 126, 141–146
Titanium carbide 145–6
Titanium dioxide 140, 145–6
Titanium nitride 142, 145, 146
Titanium oxide 143
Titanium tetrachloride 140, 142, 143
Titanium-nickel alloy 145
TNF – see tumour necrosis factor
TNO 100, 115
Toluene 32
Toluene sulphonamide formaldehyde 32
Tooth enamel 27
Toothbrushes 26
Toothpaste 27, 30, 73, 82
Toyota 128
Traffic lights 126, 147
Transmission electron microscopy 78
Trehala mana 102
Trehalose 97, 101–105
Trichloroethylene 62
Triclosan 41, 81, 82, 216
Triethylene glycol 174

Trifluoracetic acid 63
Trimethyl gallium 151
Trimethylamine 77
Truffle 69
Tuberculosis 46
Tumour necrosis factor 45, 52, 53
Tuohy, Kevin 22
Turmeric 119, 122
Turner 182, 190
Typhoid 59
Tyrian purple 188

u
Ulcers 49, 115
Ultramarine 178, 184
Ultraviolet light, effects of 24, 33, 166–7, 180
Urea 78, 217
Urea peroxide 28
Urea-aldehyde resin 198
Urine 69, 173
US Centers for Disease Control 34
US Materials Research Society 28
US National Cancer Institute 14, 81
UV irradiation 24, 45

v
Vaccines 105
Vagina 84
Van Aalten, Daan 55
Van Dyck 181
Van Gogh 204
Van Mater, H.L. 79
Vane, John 50
Vanilloid receptor type 1 120
Varnishes used by artists 190
Vaseline 43
Vasodilator 18
Velázquez 181
Verdigris 188
Verez-Bencomo, Vicente 58
Vermeer 202
Vermilion 183, 196
Verney, Ralph 26
Vikesland, Peter 82
Vinyl chloride 62
Vioxx 51–53
Viridian 203
Viruses 98
Visible spectrum of dyes 180
Vitamins 40, 45, 119
Vitiligo 38
VNO – see vomero-nasal organ
Vomero-nasal organ 71

w
Wallace, John H. 79
Wang Yuan-lu 187
Washing soda 161
Washington, George 26
Washing-up liquids 160
Water glass 162
Water hardness 162–3
Water softener 169–70
Wedzicha, Bronislaw 124
Wells, Horace 61
Wesley Jessen 20, 22
Wessel, Gary 56
Wheat 116
White pigments 189
Whitfield, George 135
WHO – see World Health Organisation
Wichterle, Otto 22
Window cleaning 140
Woad 184
Wolbers, Richard 198
Wood ash 156
Wool, washing 156
World Health Organisation 104, 106, 124
Wysocki, Charles 71

x
XCell 60
Xenon 63–4
X-ray fluorescence 193, 200
Xylene solvent 178
Xylos 60

y
Yakult 115
Yamagishi, Kazu 30
Yanagaida, Shozo 133
Yang, Peidong 151
Yellow pigments 186–7
Yodkin, John 96
Yogurt 97

z
Zeolites 162

Zhang, Weiya 49
Zinc 77, 131
Zinc acetate 170
Zinc carbonate 170
Zinc oxide 81, 151
Zinc phenolsulfonate 81
Zinc ricinoleate 175
Zirconium chlorohydrate 79
Zirconium salts 75, 78, 80
Zooplankton 162

Related Titels

Arthur Greenberg
From Alchemie to Chemistry in Picture and Story
2007, ISBN-13: 978-0-471-75154-0

Herbert W. Roesky
Spectacular Chemical Experiments
(Foreword by George A. Olah)
2007, ISBN-13: 978-3-527-31865-0

Steven A. Edwards
The Nanotech Pioneers
Where Are They Taking Us?
2006, ISBN: 978-3-527-31290-0

Jürgen Audretsch
Entangled World
The Fascination of Quantum Information and Computation
2005, ISBN-13: 978-3-527-40470-4

Jürgen Renn
Albert Einstein – Chief Engineer of the Universe
One Hundred Authors for Einstein
2005, ISBN-13: 978-3-527-40574-9

Reiner Braun, David Krieger
Einstein – Peace Now!
Visions and Ideas
2005, ISBN-13: 978-3-527-40604-3

Hubertus P. Bell, Tim Feuerstein, Carlos E. Günter, Sören Hölsken, Jan Klaas Lohmann
What's Cooking in Chemistry?
How Leading Chemists Succeed in the Kitchen
2003, ISBN-13: 978-3-527-30723-4

Carl Djerassi, Roald Hoffmann
Oxygen
A Play in 2 Acts
2001, ISBN-13: 978-3-527-30413-4